AWS CLOUD AUTOMATION

HARNESSING TERRAFORM FOR AWS INFRASTRUCTURE AS CODE

4 BOOKS IN 1

BOOK 1
AWS CLOUD AUTOMATION: TERRAFORM ESSENTIALS FOR BEGINNERS

BOOK 2
MASTERING TERRAFORM: ADVANCED TECHNIQUES FOR AWS CLOUD AUTOMATION

BOOK 3
OPTIMIZING AWS INFRASTRUCTURE: ADVANCED TERRAFORM STRATEGIES

BOOK 4
EXPERT AWS CLOUD AUTOMATION: SCALING AND MANAGING COMPLEX DEPLOYMENTS WITH TERRAFORM

ROB BOTWRIGHT

Published by Rob Botwright
Library of Congress Cataloging-in-Publication Data
ISBN 978-1-83938-707-4
Cover design by Rizzo

Disclaimer

The contents of this book are based on extensive research and the best available historical sources. However, the author and publisher make no claims, promises, or guarantees about the accuracy, completeness, or adequacy of the information contained herein. The information in this book is provided on an "as is" basis, and the author and publisher disclaim any and all liability for any errors, omissions, or inaccuracies in the information or for any actions taken in reliance on such information. The opinions and views expressed in this book are those of the author and do not necessarily reflect the official policy or position of any organization or individual mentioned in this book. Any reference to specific people, places, or events is intended only to provide historical context and is not intended to defame or malign any group, individual, or entity. The information in this book is intended for educational and entertainment purposes only. It is not intended to be a substitute for professional advice or judgment. Readers are encouraged to conduct their own research and to seek professional advice where appropriate. Every effort has been made to obtain necessary permissions and acknowledgments for all images and other copyrighted material used in this book. Any errors or omissions in this regard are unintentional, and the author and publisher will correct them in future editions.

BOOK 1 - AWS CLOUD AUTOMATION: TERRAFORM ESSENTIALS FOR BEGINNERS

BOOK 2 - MASTERING TERRAFORM: ADVANCED TECHNIQUES FOR AWS CLOUD AUTOMATION

BOOK 3 - OPTIMIZING AWS INFRASTRUCTURE: ADVANCED TERRAFORM STRATEGIES

BOOK 4 - EXPERT AWS CLOUD AUTOMATION: SCALING AND MANAGING COMPLEX DEPLOYMENTS WITH TERRAFORM

Introduction

Welcome to "Harnessing Terraform for AWS Infrastructure as Code," a comprehensive book bundle designed to equip you with the knowledge and skills needed to master Terraform for automating and managing your AWS infrastructure. This bundle consists of four books, each tailored to address different levels of expertise and covering various aspects of Terraform usage on the AWS cloud.

Book 1, "AWS Cloud Automation: Terraform Essentials for Beginners," serves as your entry point into the world of Terraform. Whether you're new to infrastructure as code or just getting started with Terraform, this book will guide you through the essential concepts and provide hands-on tutorials to help you become proficient in defining, provisioning, and managing AWS resources using Terraform.

Once you've grasped the basics, Book 2, "Mastering Terraform: Advanced Techniques for AWS Cloud Automation," takes you deeper into Terraform's advanced features and capabilities. From managing state and dependencies to implementing modularization and reusable modules, this book equips you with the skills needed to tackle more complex infrastructure automation tasks with confidence.

Book 3, "Optimizing AWS Infrastructure: Advanced Terraform Strategies," focuses on optimizing your AWS infrastructure deployments using Terraform. Learn how to minimize costs, enhance scalability, and improve resource utilization through optimization techniques and best practices, ensuring your infrastructure meets evolving business requirements efficiently.

Finally, Book 4, "Expert AWS Cloud Automation: Scaling and Managing Complex Deployments with Terraform," provides advanced insights into Terraform's capabilities for scaling and managing complex AWS deployments. Dive into topics such as orchestrating multi-region architectures, implementing advanced networking configurations, and handling sophisticated deployment workflows with ease.

Whether you're a beginner looking to build a strong foundation or an experienced practitioner seeking to refine your skills, this book bundle has something for everyone. By the end of this journey, you'll be well-equipped to harness the power of Terraform for AWS infrastructure as code and drive innovation and efficiency in your organization's cloud environment. Let's embark on this exciting journey together!

BOOK 1
AWS CLOUD AUTOMATION
TERRAFORM ESSENTIALS FOR BEGINNERS

ROB BOTWRIGHT

Chapter 1: Introduction to AWS Cloud and Infrastructure as Code

AWS offers a vast array of services designed to cater to various computing needs, ranging from computing power to storage, databases, machine learning, and beyond. Understanding the breadth and depth of AWS services is essential for efficiently architecting and deploying applications in the cloud. One of the core services provided by AWS is Amazon Elastic Compute Cloud (EC2), which offers resizable compute capacity in the cloud. To provision an EC2 instance using the AWS CLI, you can use the **aws ec2 run-instances** command, specifying parameters such as the instance type, AMI, and security group. Another fundamental service is Amazon Simple Storage Service (S3), which provides scalable object storage for data backup, archiving, and analytics. To create an S3 bucket using the AWS CLI, you can use the **aws s3 mb s3://bucket-name** command, replacing "bucket-name" with your desired bucket name.

AWS also offers managed database services like Amazon Relational Database Service (RDS), which supports various database engines such as MySQL, PostgreSQL, and Amazon Aurora. Deploying an RDS instance can be done through the AWS Management Console or using the AWS CLI with commands like **aws rds create-db-instance**. For developers looking to build serverless applications, AWS Lambda provides a compute service that runs code in response to events and automatically scales as needed. To create a Lambda function using the AWS CLI, you can use the **aws lambda create-function** command, specifying the runtime, handler, and other configuration options.

In addition to compute and storage services, AWS offers a wide range of tools for developers and IT professionals to manage and monitor their infrastructure. AWS CloudFormation allows users to define infrastructure as code using a template format, enabling automated provisioning and management of AWS resources. To deploy a CloudFormation stack using the AWS CLI, you can use the **aws cloudformation create-stack** command, providing the stack name and template file as arguments. AWS Identity and Access Management (IAM) enables granular control over user permissions and access to AWS resources. With the AWS CLI, you can create IAM users, groups, and policies using commands like **aws iam create-user** and **aws iam create-policy**.

For organizations seeking to enhance their security posture, AWS offers services like Amazon GuardDuty, a managed threat detection service that continuously monitors for malicious activity and unauthorized behavior. Setting up GuardDuty can be accomplished through the AWS Management Console, where users can enable the service and configure findings to be sent to CloudWatch or S3 for further analysis. Moreover, AWS Config provides a detailed inventory of AWS resources and configuration changes, helping organizations assess compliance and track resource relationships over time. To enable AWS Config with the AWS CLI, you can use the **aws configservice put-configuration-recorder** command, specifying the desired configuration recorder settings.

AWS also offers a range of machine learning services, including Amazon SageMaker, a fully managed service for building, training, and deploying machine learning models at scale. Developers can use the AWS CLI to create SageMaker notebook instances, training jobs, and endpoints using commands like **aws sagemaker create-notebook-instance** and **aws sagemaker create-training-job**. Additionally, Amazon Polly and Amazon Rekognition provide capabilities for text-to-speech conversion and image and video analysis, respectively, allowing developers to integrate advanced AI functionalities into their applications with ease.

In summary, AWS offers a comprehensive suite of services and tools designed to meet the diverse needs of modern businesses and developers. From compute and storage to machine learning and security, AWS provides the building blocks necessary to architect scalable, resilient, and secure cloud-based solutions. By leveraging the power of AWS services and understanding how to deploy them effectively using the AWS CLI, organizations can accelerate innovation, reduce time to market, and drive business success in the cloud era.

Infrastructure as Code (IaC) is a transformative approach to managing IT infrastructure, enabling organizations to automate the provisioning and configuration of resources using code. This methodology offers numerous benefits that streamline operations, enhance scalability, improve reliability, and promote collaboration across development and operations teams. One of the key advantages of IaC is its ability to increase the speed and agility of infrastructure deployment through automation. By defining infrastructure configurations in code, organizations can rapidly provision resources, replicate environments, and scale infrastructure to meet evolving business demands. AWS CloudFormation is a prominent IaC service that allows users to define infrastructure as code using a template format, automating the deployment of AWS resources with a single command. To deploy a CloudFormation stack, developers can use the **aws cloudformation create-stack** command, specifying the stack name and template file as arguments.

Another benefit of IaC is improved consistency and reliability across environments. Traditional manual processes for provisioning and configuring infrastructure are prone to human error and inconsistencies, leading to configuration drift and potential downtime. With IaC, infrastructure configurations are codified and version-controlled, ensuring that deployments are consistent and reproducible across development, testing, and production environments. Tools like Terraform provide a declarative language for defining infrastructure configurations, enabling users to manage resources across multiple cloud providers with a unified workflow. Deploying infrastructure with Terraform involves writing configuration files in HashiCorp Configuration Language (HCL) and executing commands like **terraform init**, **terraform plan**, and **terraform apply** to initialize the project, preview changes, and apply configurations, respectively.

Additionally, IaC facilitates better collaboration and alignment between development and operations teams by codifying infrastructure requirements and dependencies. By treating infrastructure as code, developers and operations engineers can work together to define infrastructure configurations, automate deployments, and integrate infrastructure changes into continuous integration and continuous delivery (CI/CD) pipelines. This collaboration fosters a culture of shared responsibility and accountability, where teams can leverage version control systems like Git to track changes, review code, and collaborate on infrastructure improvements. Moreover, IaC enables organizations to implement infrastructure policies and governance controls as code, ensuring compliance with security and regulatory requirements. AWS Identity and Access Management (IAM) policies, for example, can be defined using JSON or YAML syntax and deployed using the **aws iam put-policy** command, allowing organizations to manage permissions and access controls programmatically.

Furthermore, IaC promotes infrastructure automation and repeatability, reducing the time and effort required to deploy and manage complex environments. By codifying infrastructure configurations, organizations can create reusable templates and modules that standardize deployment patterns and simplify the provisioning of resources. This automation not only accelerates time to market but also minimizes manual intervention and human error, resulting in more reliable and predictable infrastructure deployments. With AWS Elastic Beanstalk, developers can deploy and manage web applications and services at scale with ease, leveraging preconfigured environment templates and automation features. Deploying an application with Elastic Beanstalk involves creating an application source bundle, defining environment configurations in a YAML or JSON file, and using the **eb create** command to launch the environment.

Moreover, IaC enables organizations to embrace infrastructure evolution and innovation by empowering teams to experiment, iterate, and adapt infrastructure configurations as requirements change. By leveraging version control systems and infrastructure as code practices, organizations can implement feedback loops and continuous improvement processes that drive innovation and agility. AWS CodePipeline, for instance, enables users to automate the build, test, and deployment phases of their application delivery process, integrating with services like AWS CodeBuild and AWS CodeDeploy to streamline CI/CD workflows. Deploying a pipeline with CodePipeline involves defining pipeline configurations in a JSON or YAML file and using the AWS Management Console or AWS CLI to create the pipeline.

In summary, the benefits of infrastructure as code are manifold, offering organizations a powerful framework for automating and managing cloud infrastructure. By treating infrastructure as code, organizations can accelerate deployment velocity, improve consistency and reliability, foster collaboration and alignment, enforce policies and governance controls, and drive innovation and agility. With a robust set of tools and services available from cloud providers like AWS, organizations can leverage infrastructure as code to optimize their operations, reduce costs, and stay competitive in today's rapidly evolving digital landscape.

Chapter 2: Getting Started with Terraform: Installation and Setup

Installing Terraform is the initial step towards leveraging its capabilities for infrastructure provisioning and management. Terraform, developed by HashiCorp, is an open-source tool that enables users to define and provision infrastructure as code. Before getting started with Terraform, it's essential to install the tool on your local machine or a server where you plan to manage your infrastructure. The installation process varies depending on your operating system, but HashiCorp provides official installation packages and binaries for Windows, macOS, and Linux distributions.

For users on Windows, installing Terraform involves downloading the Terraform executable and adding it to your system's PATH environment variable. To download Terraform, you can visit the official Terraform website or use a package manager like Chocolatey. Once downloaded, extract the Terraform executable from the ZIP archive and move it to a directory included in your system's PATH. You can then verify the installation by opening a command prompt and running the **terraform --version** command, which should display the installed Terraform version.

Similarly, on macOS, you can install Terraform using a package manager like Homebrew or by downloading the Terraform binary directly from the HashiCorp website. If using Homebrew, you can run the **brew**

install terraform command to install Terraform and then verify the installation by running **terraform -- version** in the terminal. Alternatively, you can download the Terraform binary, extract it, and move it to a directory in your system's PATH.

For Linux users, installing Terraform typically involves downloading the Terraform binary and placing it in a directory included in the system's PATH. You can use tools like wget or curl to download the Terraform binary from the HashiCorp website. Once downloaded, extract the binary and move it to a location such as **/usr/local/bin**. You can then verify the installation by running **terraform --version** in the terminal.

Alternatively, if you prefer to use package managers on Linux, HashiCorp provides official Terraform packages for popular distributions such as Ubuntu, CentOS, and Debian. You can add the HashiCorp GPG key to your system, configure the package repository, and then install Terraform using the package manager's installation command. For example, on Ubuntu, you can run the following commands:

bashCopy code

curl -fsSL https://apt.releases.hashicorp.com/gpg | sudo apt-key add - sudo apt-add-repository "deb [arch=amd64] https://apt.releases.hashicorp.com $(lsb_release -cs) main" sudo apt-get update && sudo apt-get install terraform

After installing Terraform, it's crucial to verify that the installation was successful and that Terraform is accessible from the command line. You can do this by running the **terraform --version** command, which should display the installed Terraform version without any errors. Additionally, you can run **terraform** without any arguments to see a list of available commands and options, confirming that Terraform is installed and configured correctly.

In summary, installing Terraform is a straightforward process that involves downloading the Terraform binary or package for your operating system, adding it to your system's PATH, and verifying the installation by running **terraform --version**. Once installed, you can begin using Terraform to define, provision, and manage your infrastructure as code, enabling automation, scalability, and consistency in your cloud environment.

Configuring AWS credentials is a crucial step for interacting with AWS services programmatically or through command-line tools such as the AWS Command Line Interface (CLI) or software development kits (SDKs) for various programming languages. AWS employs a secure authentication mechanism based on access keys, consisting of an Access Key ID and a Secret Access Key, which are used to authenticate requests to AWS services. To configure AWS credentials, you can use the AWS Management Console, environment variables, or configuration files.

One common method for configuring AWS credentials is using environment variables. This approach is convenient for temporary or ad-hoc use cases, such as running commands in a terminal session. To configure AWS credentials using environment variables, you need to set two variables: **AWS_ACCESS_KEY_ID** and **AWS_SECRET_ACCESS_KEY**, which correspond to your access key ID and secret access key, respectively. Additionally, you can optionally set the **AWS_DEFAULT_REGION** variable to specify the default AWS region for API requests. For example, on Unix-based systems like Linux or macOS, you can run the following commands in your terminal:

bashCopy code

export AWS_ACCESS_KEY_ID=your-access-key-id export AWS_SECRET_ACCESS_KEY=your-secret-access-key export AWS_DEFAULT_REGION=us-east-1

On Windows, you can use the **set** command to set environment variables:

batchCopy code

set AWS_ACCESS_KEY_ID=your-access-key-id set AWS_SECRET_ACCESS_KEY=your-secret-access-key set AWS_DEFAULT_REGION=us-east-1

Alternatively, you can configure AWS credentials using the AWS CLI, which provides a **configure** command to interactively set up credentials and default settings. To configure AWS credentials with the AWS CLI, you can run the following command and follow the prompts:

bashCopy code

aws configure

This command will prompt you to enter your access key ID, secret access key, default region, and default output format (e.g., JSON). Once entered, the AWS CLI will store these credentials in a configuration file located in your home directory (**~/.aws/credentials** on Unix-based systems or **%UserProfile%\.aws\credentials** on Windows). These credentials will be used by default for subsequent AWS CLI commands unless overridden by environment variables or command-line options.

Another method for configuring AWS credentials is using AWS Identity and Access Management (IAM) roles. IAM roles provide temporary security credentials that applications or services can use to make requests to AWS services on behalf of users or resources. IAM roles are particularly useful for running applications or services on AWS infrastructure, as they eliminate the need to manage long-term access keys and secrets. To configure an IAM role, you can use the AWS Management Console or the AWS CLI to create a role with the necessary permissions and then assign the role to your EC2 instance or other AWS resource. For example, using the AWS CLI, you can create an IAM role with the **create-role** command:

bashCopy code

```
aws iam create-role --role-name my-role --assume-role-policy-document file://trust-policy.json
```

In this command, **trust-policy.json** is a JSON file containing the trust policy document that specifies which AWS entities are allowed to assume the role. Once the role is created, you can attach policies to it using the **attach-role-policy** command:

bashCopy code

```
aws iam attach-role-policy --role-name my-role --policy-arn arn:aws:iam::aws:policy/AmazonS3ReadOnlyAccess
```

This command attaches the **AmazonS3ReadOnlyAccess** policy to the **my-role** role, granting it read-only access to Amazon S3 resources. Finally, you can assign the role to your EC2 instance using the **associate-iam-instance-profile** command:

bashCopy code

aws ec2 associate-iam-instance-profile --instance-id i-1234567890abcdef0 --iam-instance-profile Name=my-role

This command associates the **my-role** IAM instance profile with the specified EC2 instance, allowing the instance to assume the role and access AWS services according to the role's permissions.

In summary, configuring AWS credentials is an essential step for interacting with AWS services securely and programmatically. Whether using environment variables, the AWS CLI, or IAM roles, properly configured credentials are necessary for authenticating requests to AWS services and accessing resources in your AWS environment. By following best practices for credential management and security, you can ensure that your applications and services operate smoothly and securely in the AWS cloud.

Chapter 3: Understanding Terraform Configuration Language

Declarative and imperative configuration are two contrasting approaches to defining and managing infrastructure and application resources. Each methodology has its own advantages and use cases, and understanding the differences between them is essential for making informed decisions when architecting and deploying systems in cloud environments. Declarative configuration focuses on describing the desired state of the system without specifying how to achieve that state, while imperative configuration involves specifying step-by-step instructions for achieving the desired state.

In a declarative configuration approach, users define the desired end state of the system, and the underlying infrastructure or application management tool determines the actions required to achieve that state. This approach emphasizes what should be done rather than how it should be done, allowing for a more abstract and high-level representation of system configurations. One of the primary benefits of declarative configuration is its simplicity and ease of use, as users only need to specify the desired configuration without worrying about the implementation details. Additionally, declarative configuration promotes idempotent behavior, meaning that applying the configuration multiple times results in the same end state, regardless of the current state of the system.

A common example of declarative configuration is the use of configuration management tools like Puppet, Chef, or Ansible, which allow users to define system configurations using descriptive language constructs such as manifests, playbooks, or roles. These tools use idempotent operations to ensure that the system configuration remains consistent with the desired state, even in the face of changes or failures. For example, with Ansible, users can define tasks and roles in YAML files, specifying the desired configuration of servers, services, and applications. Deploying an Ansible playbook involves running the **ansible-playbook** command with the path to the playbook file as an argument, such as:

bashCopy code

```
ansible-playbook playbook.yml
```

This command instructs Ansible to apply the configuration defined in the **playbook.yml** file to the target hosts, ensuring that the desired state is achieved across the infrastructure.

In contrast, imperative configuration focuses on specifying the exact steps or commands required to achieve a desired state, rather than describing the end state itself. This approach is more procedural and prescriptive, as users must explicitly define each action or operation needed to configure the system. Imperative configuration is often used in imperative programming languages or scripting environments, where users write scripts or code that directly manipulate the system's state. While imperative configuration provides more control and flexibility over the configuration process, it can be more complex and error-prone, as users must handle edge cases and manage the system's state explicitly.

An example of imperative configuration is the use of shell scripts or command-line tools to provision and configure infrastructure resources. For instance, to create an EC2 instance in AWS using the AWS CLI, users must specify all the necessary parameters and options required to provision the instance, such as the instance type, AMI, and security groups. The **aws ec2 run-instances** command is used to launch EC2 instances, with options like **--instance-type**, **--image-id**, and **--security-group-ids** specifying the configuration details. For example:

bashCopy code

aws ec2 run-instances --instance-type t2.micro --image-id ami-1234567890abcdef0 --security-group-ids sg-1234567890abcdef0

This command creates a new EC2 instance with the specified instance type, AMI, and security groups, following a step-by-step imperative approach to provisioning the resource.

In summary, declarative and imperative configuration are two distinct paradigms for defining and managing system configurations. Declarative configuration emphasizes the desired end state of the system, while imperative configuration focuses on the specific steps or commands required to achieve that state. Each approach has its own strengths and weaknesses, and the choice between them depends on factors such as the complexity of the system, the level of control required, and the preferences of the users or development team. By understanding the differences between declarative and imperative configuration, users can select the most appropriate approach for their specific use cases and effectively manage their infrastructure and applications in cloud environments.

Terraform's resource blocks are fundamental constructs used to define and manage infrastructure resources in a declarative manner. These resource blocks encapsulate the configuration details of a specific resource type, such as an EC2 instance, a VPC, or a Route 53 record, and provide a standardized way to represent infrastructure components as code. Resource blocks are defined within Terraform configuration files using the **resource** keyword followed by the resource type and a unique resource identifier.
For example, to define an AWS EC2 instance using Terraform, you would create a resource block like this:
hclCopy code
resource "aws_instance" "example" { ami = "ami-12345678" instance_type = "t2.micro" }
In this example, **aws_instance** is the resource type, and **"example"** is the resource identifier, which can be any valid identifier name. Within the resource block, you can specify various configuration attributes and settings specific to the resource type, such as the Amazon Machine Image (AMI) ID, instance type, key pair, security groups, and so on.

Resource blocks can also reference other resources, enabling dependency management and ensuring proper ordering of resource creation and destruction. For example, you can specify dependencies between resources using the **depends_on** attribute:
hclCopy code

```
resource "aws_instance" "web_server" { ami = "ami-12345678" instance_type = "t2.micro" } resource "aws_security_group" "web_sg" { name = "web_sg" description = "Security group for web servers" # Reference the `aws_instance` resource by its unique identifier depends_on = [aws_instance.web_server] # Define security group rules... }
```

In this example, the **aws_security_group** resource depends on the **aws_instance** resource named **"web_server"**, ensuring that the security group is created only after the EC2 instance has been provisioned.

Furthermore, Terraform's resource blocks support interpolation syntax, allowing you to dynamically reference values from other resources or data sources. Interpolation expressions are enclosed in **${}** and can be used within attribute values to inject runtime values or perform calculations. For example, you can reference attributes of other resources like this:
hclCopy code

```
resource "aws_instance" "web_server" { ami = "ami-12345678" instance_type = "t2.micro" } resource
```

"aws_eip" "web_server_ip" { instance = aws_instance.web_server.id }

In this example, the Elastic IP (**aws_eip**) resource's **instance** attribute references the ID of the EC2 instance (**aws_instance.web_server**) using interpolation syntax, ensuring that the Elastic IP is associated with the correct instance.

Moreover, Terraform's resource blocks support meta-arguments such as **count, for_each**, and **provider**, which provide additional control over resource creation and configuration. The **count** meta-argument allows you to create multiple instances of a resource based on a numeric value or a list, while the **for_each** meta-argument enables dynamic resource creation based on a map or set of strings.

For instance, you can use the **count** meta-argument to create multiple EC2 instances with different instance types:

```hcl
variable "instance_types" { default = ["t2.micro", "t2.small", "t2.medium"] } resource "aws_instance" "web_servers" { count = length(var.instance_types) ami = "ami-12345678" instance_type = var.instance_types[count.index] }
```

In this example, the **aws_instance** resource will be instantiated three times, each with a different instance type specified by the **instance_types** variable.

Furthermore, Terraform's resource blocks support the **provider** meta-argument, allowing you to specify the provider configuration for a particular resource. This is useful when working with multiple providers or when overriding default provider configurations.

For example, you can define an AWS EC2 instance using a specific AWS provider configuration:

hclCopy code

```
provider "aws" { region = "us-east-1" } resource "aws_instance" "web_server" { provider = aws.east ami = "ami-12345678" instance_type = "t2.micro" }
```

In this example, the **aws_instance** resource uses the **aws.east** provider configuration to provision the EC2 instance in the **us-east-1** region.

In summary, Terraform's resource blocks are powerful constructs that enable infrastructure-as-code practitioners to define, configure, and manage cloud resources declaratively. Resource blocks provide a standardized and expressive way to represent infrastructure components as code, facilitating infrastructure provisioning, dependency management, and dynamic resource creation. By leveraging Terraform's resource blocks and associated features such as interpolation syntax, meta-arguments, and provider configurations, users can efficiently define and manage complex infrastructure deployments in a scalable and maintainable manner.

Chapter 4: Creating and Managing AWS Resources with Terraform

Provisioning EC2 instances is a fundamental task in cloud computing, allowing users to deploy virtual servers on the Amazon Web Services (AWS) platform to run their applications and workloads. AWS EC2 instances provide scalable compute capacity in the cloud, enabling users to quickly launch, resize, and manage virtual servers to meet their computing needs. Provisioning EC2 instances can be done using various methods, including the AWS Management Console, AWS Command Line Interface (CLI), or infrastructure-as-code tools like Terraform.

Using the AWS Management Console is one of the simplest ways to provision EC2 instances, providing a graphical user interface (GUI) for users to interactively configure and launch virtual servers. To provision an EC2 instance using the AWS Management Console, users can navigate to the EC2 service dashboard, click on the "Launch Instance" button, and follow the step-by-step wizard to configure instance settings such as instance type, AMI (Amazon Machine Image), instance details, storage, security groups, and key pairs. Once all configurations are set, users can review their selections and launch the instance. This process guides users through the various options available for

EC2 instance provisioning, making it accessible to users who prefer a visual interface.

Alternatively, provisioning EC2 instances can be automated using the AWS Command Line Interface (CLI), which provides a set of commands for interacting with AWS services from the command line. To provision an EC2 instance using the AWS CLI, users need to first configure their AWS credentials using the **aws configure** command, providing their access key ID, secret access key, default region, and output format. Once configured, users can use the **aws ec2 run-instances** command to launch EC2 instances with specified parameters such as instance type, AMI ID, key pair, security groups, and tags. For example, to launch a t2.micro instance with the Amazon Linux 2 AMI, users can run the following command:

bashCopy code

```
aws ec2 run-instances --image-id ami-1234567890abcdef0 --instance-type t2.micro --key-name MyKeyPair --security-group-ids sg-1234567890abcdef0
```

This command creates a new EC2 instance based on the specified AMI, instance type, key pair, and security group.

Moreover, infrastructure-as-code tools like Terraform provide a declarative approach to provisioning and managing infrastructure resources, including EC2 instances. With Terraform, users define infrastructure configurations in code using a domain-specific language (DSL) called HashiCorp Configuration Language (HCL) or JSON. To provision EC2 instances using Terraform, users create a Terraform configuration file (e.g., **main.tf**) and define an **aws_instance** resource block with the desired configuration settings, such as instance type, AMI, key pair, and security groups. For example, a Terraform configuration to provision an EC2 instance might look like this:

```
hclCopy code
resource "aws_instance" "my_instance" { ami = "ami-1234567890abcdef0" instance_type = "t2.micro" key_name = "MyKeyPair" security_groups = ["MySecurityGroup"] }
```

After defining the Terraform configuration, users initialize the Terraform project using the **terraform init** command, which downloads the necessary provider plugins and modules. Then, they can run the **terraform plan** command to preview the changes Terraform will make to the infrastructure, followed by the **terraform apply** command to apply the changes and provision the EC2 instance. Terraform manages the lifecycle of the infrastructure resources, ensuring that the desired state defined in the configuration is realized in the AWS environment.

In addition to the AWS Management Console, AWS CLI, and Terraform, users can also provision EC2 instances programmatically using AWS SDKs

(Software Development Kits) for various programming languages, such as Python, Java, JavaScript, and .NET. These SDKs provide language-specific APIs for interacting with AWS services, allowing developers to integrate EC2 instance provisioning into their applications and workflows. By leveraging the AWS SDKs, users can programmatically launch EC2 instances, customize instance configurations, and automate infrastructure provisioning tasks as part of their software development processes.

In summary, provisioning EC2 instances is a fundamental task in cloud computing, enabling users to deploy virtual servers on the AWS platform to run their applications and workloads. Whether using the AWS Management Console, AWS CLI, Terraform, or AWS SDKs, users have multiple options for provisioning EC2 instances to meet their specific requirements and preferences. Each method provides a different level of control and automation, allowing users to choose the approach that best suits their use case and workflow. With these provisioning tools and techniques, users can efficiently deploy and manage EC2 instances in the AWS cloud, supporting their diverse computing needs and business objectives.

Managing AWS networking resources is a crucial aspect of building and operating applications in the cloud, as it involves configuring and controlling the network infrastructure that enables communication between various components and services within an

AWS environment. AWS offers a comprehensive set of networking services and features to help users design, deploy, and manage scalable and secure network architectures. These networking resources include virtual private clouds (VPCs), subnets, route tables, internet gateways, virtual private gateways, elastic IP addresses, security groups, network access control lists (NACLs), and more. Managing these networking resources involves tasks such as creating and configuring VPCs, defining subnets, setting up routing tables, configuring network security, and establishing connectivity to the internet and other networks.

One of the fundamental networking resources in AWS is the Virtual Private Cloud (VPC), which provides users with a logically isolated section of the AWS cloud where they can launch resources in a virtual network. To manage VPCs using the AWS Management Console, users can navigate to the VPC dashboard, click on the "Create VPC" button, and specify the VPC's name, CIDR block, and other configuration settings. Alternatively, users can use the AWS CLI to create a VPC using the **create-vpc** command, specifying the desired parameters such as the CIDR block. For example:

bashCopy code

aws ec2 create-vpc --cidr-block 10.0.0.0/16

This command creates a new VPC with the specified CIDR block.

Once a VPC is created, users can further customize its configuration by creating subnets, route tables, and internet gateways. Subnets allow users to partition the VPC's IP address range into smaller segments and deploy resources within specific availability zones. To create a subnet using the AWS CLI, users can run the **create-subnet** command, specifying the VPC ID, CIDR block, and availability zone. For example:

```
bashCopy code
aws ec2 create-subnet --vpc-id vpc-1234567890abcdef0 --cidr-block 10.0.1.0/24 --availability-zone us-east-1a
```

This command creates a subnet within the specified VPC and availability zone.

Route tables control the routing of network traffic within a VPC, specifying the destination for traffic based on its destination IP address. Users can create custom route tables and associate them with subnets to define how traffic should be routed. To create a route table using the AWS CLI, users can run the **create-route-table** command, specifying the VPC ID. For example:

```
bashCopy code
aws ec2 create-route-table --vpc-id vpc-1234567890abcdef0
```

This command creates a new route table within the specified VPC.

Internet gateways provide a way for resources within a VPC to communicate with the internet and other

AWS services. Users can attach an internet gateway to a VPC and configure route tables to direct internet-bound traffic to the gateway. To create an internet gateway using the AWS CLI, users can run the **create-internet-gateway** command. For example:

bashCopy code

```
aws ec2 create-internet-gateway
```

This command creates a new internet gateway that can be attached to a VPC.

Additionally, managing AWS networking resources involves configuring network security to protect resources from unauthorized access. Security groups act as virtual firewalls that control inbound and outbound traffic for EC2 instances and other resources within a VPC. Users can create and configure security groups to define allowed traffic flows based on protocols, ports, and IP addresses. To create a security group using the AWS CLI, users can run the **create-security-group** command, specifying the VPC ID and desired inbound and outbound rules. For example:

bashCopy code

```
aws ec2 create-security-group --group-name my-security-group --description "My security group" --vpc-id vpc-1234567890abcdef0
```

This command creates a new security group within the specified VPC.

In addition to security groups, users can also configure network access control lists (NACLs) to control traffic at the subnet level. NACLs act as stateless firewalls that filter traffic based on rules defined for inbound and outbound traffic. Users can create and configure NACLs to allow or deny specific types of traffic based on IP addresses, protocols, and ports. To create a NACL using the AWS CLI, users can run the **create-network-acl** command, specifying the VPC ID. For example:

bashCopy code

```
aws    ec2    create-network-acl    --vpc-id    vpc-1234567890abcdef0
```

This command creates a new NACL within the specified VPC.

Overall, managing AWS networking resources involves a combination of creating and configuring VPCs, subnets, route tables, internet gateways, security groups, and network access control lists to design and implement scalable, secure, and highly available network architectures in the cloud. By leveraging the AWS Management Console, AWS CLI, and infrastructure-as-code tools like Terraform, users can efficiently deploy and manage their AWS networking resources to meet their specific requirements and achieve their desired network configurations.

Chapter 5: Terraform Modules: Organizing and Reusing Infrastructure Code

Structuring Terraform code with modules is a best practice for organizing and managing infrastructure configurations in a scalable and maintainable way. Terraform modules are self-contained units of configuration that encapsulate reusable components of infrastructure, allowing users to abstract complex configurations into manageable and reusable components. Modules promote code reusability, modularity, and consistency across Terraform projects, making it easier to manage infrastructure at scale and collaborate with team members. By structuring Terraform code with modules, users can improve code maintainability, simplify infrastructure provisioning, and enable better code organization and management.

To create a Terraform module, users define a directory containing one or more configuration files that define the module's resources and variables. A module directory typically contains a **main.tf** file that defines the resources and configuration settings, along with optional files for variables (**variables.tf**), outputs (**outputs.tf**), and other supporting files. Users can organize module directories within their Terraform projects or in separate repositories, depending on the level of code reuse and separation desired.

Creating a Terraform module involves defining the module's inputs (variables) and outputs, which allow users to parameterize the module and capture its outputs for use in other parts of their infrastructure. Variables are defined using the **variable** block in a module's **variables.tf** file, specifying the variable name, type, and optional default value. For example, to define a variable for the instance type in a module, users can add the following to the **variables.tf** file:

hclCopy code

variable "instance_type" { type = string default = "t2.micro" }

In this example, the **instance_type** variable is defined with a default value of **t2.micro** and a type of **string**.

Once variables are defined, users can reference them within the module's **main.tf** file to parameterize resource configurations. For example, to create an EC2 instance with the specified instance type, users can use interpolation syntax to reference the **instance_type** variable:

hclCopy code

resource "aws_instance" "example" { ami = "ami-12345678" instance_type = var.instance_type }

In this example, the **instance_type** variable is referenced using **var.instance_type**, allowing users to specify the instance type when using the module.

Additionally, modules can define outputs to expose information about the resources they create, allowing users to capture and use this information in other parts of their infrastructure. Outputs are defined using the **output** block in a module's **outputs.tf** file, specifying the output name and value. For example, to output the instance ID of the EC2 instance created by the module, users can add the following to the **outputs.tf** file:

hclCopy code

output "instance_id" { value = aws_instance.example.id }

In this example, the **instance_id** output exposes the ID of the EC2 instance created by the module, allowing users to reference it in other parts of their infrastructure.

Once a module is defined, users can use it in their Terraform configurations by referencing its source location. Modules can be referenced from local file paths, Git repositories, or module registries like the Terraform Registry. To use a module from a local file path, users can specify the path to the module directory using a relative or absolute file path. For example, to use a module located in a directory named **my_module**, users can reference it in their Terraform configuration like this:

hclCopy code

module "example_module" { source = "./my_module" }

In this example, the **source** parameter specifies the path to the module directory relative to the location of the Terraform configuration file.

Alternatively, users can reference modules from Git repositories by specifying the Git repository URL and optionally a specific Git branch, tag, or commit hash. For example, to use a module hosted on GitHub, users can specify the repository URL in the **source** parameter:

hclCopy code

```
module "example_module" { source = "git::https://github.com/example/my_module.git" }
```

In this example, the **source** parameter specifies the Git repository URL for the module.

By structuring Terraform code with modules, users can modularize their infrastructure configurations, promote code reuse, and improve collaboration and consistency across Terraform projects. Modules enable users to encapsulate reusable components of infrastructure, parameterize configurations with variables, and expose outputs for use in other parts of their infrastructure. With modules, users can build scalable and maintainable infrastructure configurations in Terraform, enabling them to manage complex cloud environments more effectively.

Reusing Terraform modules across projects is a powerful strategy for promoting code reuse, standardization, and consistency in infrastructure configurations. Terraform modules encapsulate reusable components of infrastructure, such as compute instances, databases, networking resources, and security settings, allowing users to abstract complex configurations into reusable building blocks. By reusing modules across projects, users can streamline the development and management of infrastructure, reduce duplication of effort, and ensure consistency and best practices across their Terraform codebases.

To reuse a Terraform module across projects, users can either create their own modules or leverage existing modules from the Terraform Registry, Git repositories, or other module registries. Creating a reusable module involves defining the module's resources, variables, and outputs in a self-contained directory structure that can be easily referenced and reused in different projects. Users can organize their module directories within their Terraform projects or in separate repositories, depending on their preferred workflow and level of code reuse.

Once a module is defined, users can reference it in their Terraform configurations by specifying the module's source location. Modules can be referenced from local file paths, Git repositories, or module registries like the Terraform Registry. To use a module from a local file path, users can specify the path to the module directory using a relative or absolute file path. For example, to reference a module named **my_module** located in a directory named **modules**

within the project's root directory, users can specify the path like this:
hclCopy code
module "example_module" { source = "./modules/my_module" }
In this example, the **source** parameter specifies the path to the module directory relative to the location of the Terraform configuration file.
Alternatively, users can reference modules from Git repositories by specifying the repository URL and optionally a specific Git branch, tag, or commit hash. For example, to use a module hosted on GitHub, users can specify the repository URL in the **source** parameter:
hclCopy code
module "example_module" { source = "git::https://github.com/example/my_module.git" }
In this example, the **source** parameter specifies the Git repository URL for the module.
Using existing modules from the Terraform Registry is another common approach to reusing modules across projects. The Terraform Registry is a public repository of Terraform modules contributed by the community and maintained by HashiCorp. Users can search for modules in the registry using the Terraform CLI or web interface, and then reference them in their Terraform configurations using the module's namespace and name. For example, to use a module named **vpc** from the **terraform-aws-modules**

namespace in the Terraform Registry, users can specify the module like this:

hclCopy code

```
module "example_vpc" { source = "terraform-aws-modules/vpc/aws" version = "2.0.0" }
```

In this example, the **source** parameter specifies the namespace (**terraform-aws-modules**) and name (**vpc**) of the module in the Terraform Registry, along with the desired version (**2.0.0**).

By reusing Terraform modules across projects, users can benefit from code reuse, standardization, and consistency in their infrastructure configurations. Modules enable users to encapsulate reusable components of infrastructure, abstract complex configurations into reusable building blocks, and promote best practices and conventions across their Terraform codebases. With modules, users can streamline the development and management of infrastructure, reduce duplication of effort, and ensure consistent and reliable deployments across their projects.

Chapter 6: Working with Variables and Outputs in Terraform

Using variables in Terraform is a fundamental technique for parameterizing configurations, making them more flexible, reusable, and maintainable. Variables allow users to define dynamic values that can be passed into Terraform configurations, enabling users to customize infrastructure deployments based on different environments, requirements, or preferences. Terraform supports various types of variables, including input variables, output variables, environment variables, and module variables, each serving different purposes and use cases. Leveraging variables in Terraform configurations enables users to create more dynamic and adaptable infrastructure deployments, facilitating better management and automation of cloud resources.

Input variables are one of the primary types of variables in Terraform, allowing users to pass values into their configurations when running Terraform commands. Input variables are defined using the **variable** block in Terraform configuration files, specifying the variable name, type, and optional default value. For example, to define an input variable named **instance_type** with a default value of **t2.micro**, users can add the following to their Terraform configuration file:

```hcl
hclCopy code
variable "instance_type" { type = string default = "t2.micro" }
```

In this example, the **instance_type** variable is defined with a type of **string** and a default value of **t2.micro**.

Once variables are defined, users can reference them within their Terraform configurations using interpolation syntax (**${}**) or by using the **var** prefix. For example, to use the **instance_type** variable defined above when creating an EC2 instance, users can specify the instance type like this:

```hcl
hclCopy code
resource "aws_instance" "example" { ami = "ami-12345678" instance_type = var.instance_type }
```

In this example, the **instance_type** variable is referenced using the **var.instance_type** syntax, allowing users to customize the instance type when running Terraform commands.

In addition to input variables, Terraform also supports output variables, which allow users to export information about their infrastructure after it's been created. Output variables are defined using the **output** block in Terraform configuration files, specifying the output name and value. For example, to output the public IP address of an EC2 instance created by a Terraform configuration, users can add the following to their configuration file:

```hcl
hclCopy code
```

output "instance_ip" { value = aws_instance.example.public_ip }

In this example, the **instance_ip** output variable exposes the public IP address of the **aws_instance.example** resource, allowing users to reference it after Terraform applies the configuration. Furthermore, Terraform allows users to define variables using environment variables, providing a convenient way to pass values into Terraform configurations without modifying the configuration files directly. Environment variables are prefixed with **TF_VAR_** and correspond to input variables defined in the Terraform configuration. For example, to set the value of the **instance_type** variable using an environment variable, users can run the following command:

bashCopy code

export TF_VAR_instance_type="t2.large"

In this example, the **TF_VAR_instance_type** environment variable is set to **t2.large**, which will override the default value of the **instance_type** variable when running Terraform commands.

Moreover, Terraform allows users to define variables within modules, enabling users to parameterize module configurations and customize their behavior. Module variables are defined using the **variable** block within module configurations, similar to input variables in root configurations. Users can pass values into module variables when using the module in their

Terraform configurations, allowing for greater flexibility and reusability of modules.

By using variables in Terraform configurations, users can create more dynamic, flexible, and maintainable infrastructure deployments. Variables enable users to parameterize configurations, customize behavior based on different environments or requirements, and promote code reuse and consistency across Terraform projects. Whether defining input variables, output variables, environment variables, or module variables, leveraging variables in Terraform configurations empowers users to create more adaptable and scalable infrastructure deployments in the cloud.

Retrieving outputs from Terraform resources is a crucial aspect of infrastructure management and automation, as it allows users to access information about their deployed resources and use it in subsequent operations or integrations with other systems. Terraform outputs enable users to capture and export data from their infrastructure deployments, such as IP addresses, domain names, ARNs (Amazon Resource Names), and other resource attributes, making it easier to interact with and manage the deployed resources. By defining outputs in Terraform configurations, users can retrieve and use this information for various purposes, including configuration validation, application deployment, monitoring, and integration with other tools and services.

To retrieve outputs from Terraform resources, users need to define output variables in their Terraform configuration files using the **output** block. The **output** block specifies the name of the output variable and the value to be exported. For example, to output the public IP address of an EC2 instance created by a Terraform configuration, users can define the following output variable in their configuration file:

hclCopy code

output "instance_ip" { value = aws_instance.example.public_ip }

In this example, the **instance_ip** output variable is defined, and its value is set to the public IP address of the **aws_instance.example** resource.

Once output variables are defined in the Terraform configuration, users can retrieve them using the Terraform CLI after applying the configuration. The **terraform output** command is used to retrieve output values from the Terraform state file. For example, to retrieve the value of the **instance_ip** output variable, users can run the following command:

bashCopy code

terraform output instance_ip

This command will display the value of the **instance_ip** output variable, allowing users to capture and use it in subsequent operations or scripts.

Additionally, users can also retrieve output values programmatically using the Terraform SDKs (Software Development Kits) or by parsing the Terraform state file directly. The Terraform SDKs provide language-specific APIs for interacting with Terraform state files, allowing users to query output values and integrate them into their applications or automation workflows. By leveraging the Terraform SDKs, users can programmatically access output values and use them to automate infrastructure management tasks or perform custom integrations with other systems.

Furthermore, output values can be used in conjunction with other Terraform commands and features, such as input variables, module dependencies, and remote state management. For example, output values can be passed as input variables to other Terraform configurations, allowing users to reference outputs from one configuration in another configuration. This enables users to create modular and interconnected infrastructure deployments, where outputs from one module can be consumed as inputs by another module, facilitating greater flexibility and reusability of Terraform code.

Moreover, Terraform outputs can be integrated with external systems and tools using Terraform's remote state management features, such as Terraform Cloud or Terraform Enterprise. These platforms provide centralized state storage and management capabilities, allowing users to securely store and access Terraform state files, including output values, from a centralized location. By leveraging remote state management, users can share output values across teams, environments, and projects, enabling collaboration and consistency in infrastructure deployments.

In summary, retrieving outputs from Terraform resources is a critical aspect of infrastructure management and automation, allowing users to access information about their deployed resources and use it in subsequent operations or integrations with other systems. By defining output variables in Terraform configurations and using the Terraform CLI or SDKs to retrieve output values, users can capture and use data from their infrastructure deployments for various purposes, including configuration validation, application deployment, monitoring, and integration with other tools and services. Terraform outputs facilitate greater flexibility, reusability, and automation in infrastructure management, enabling users to build and manage cloud infrastructure more effectively.

Chapter 7: Managing State and Remote Backends

Terraform state management is a critical aspect of infrastructure as code (IaC) workflows, as it involves tracking and managing the state of deployed resources to ensure consistency, reliability, and maintainability of infrastructure configurations. Terraform maintains a state file that records the current state of the infrastructure managed by Terraform, including the IDs, attributes, and relationships of provisioned resources. The Terraform state file serves as the source of truth for the infrastructure managed by Terraform, enabling Terraform to track changes, perform updates, and maintain the desired state of the infrastructure. Effective state management is essential for managing infrastructure deployments at scale, ensuring that changes are applied correctly and consistently across environments.

Terraform stores its state locally by default, in a file named **terraform.tfstate** in the root directory of the Terraform configuration. While local state management is convenient for development and testing purposes, it is not suitable for production environments or collaborative workflows, as it can lead to issues with state consistency and synchronization. To address these challenges, Terraform provides several options for managing state remotely, including backend configurations and remote state storage solutions.

One of the most common approaches to remote state management in Terraform is using remote state backends, which allow Terraform to store its state data in a centralized, shared location accessible to multiple users and environments. Terraform supports various backend types, including Amazon S3, Azure Blob Storage, Google Cloud Storage, HashiCorp Consul, and HashiCorp Terraform Cloud. Configuring a remote backend in Terraform involves specifying the backend type and configuration settings in the Terraform configuration file or via CLI flags.

For example, to configure Terraform to use an Amazon S3 bucket as the remote backend, users can define the backend configuration in their Terraform configuration file (**backend.tf**) like this:

hclCopy code

terraform { backend "s3" { bucket = "my-terraform-state-bucket" key = "terraform.tfstate" region = "us-west-2" encrypt = true dynamodb_table = "terraform-state-lock" } }

In this example, the **s3** backend type is specified, along with the S3 bucket name, key prefix, AWS region, encryption settings, and optional DynamoDB table for state locking.

Once a remote backend is configured, Terraform automatically stores its state data in the specified remote location, ensuring that the state is shared and synchronized across environments and team members. Remote state management provides several benefits, including improved collaboration, centralized state storage, enhanced security, and better support for automation and continuous integration/continuous deployment (CI/CD) workflows.

Additionally, Terraform supports state locking mechanisms to prevent concurrent modifications to the Terraform state by multiple users or processes. State locking helps prevent race conditions and conflicts that can arise when multiple users attempt to apply changes to the same infrastructure simultaneously. Terraform supports state locking using various mechanisms, including file-based locks, Consul locks, and DynamoDB locks, depending on the chosen backend.

To enable state locking with a remote backend, users can specify the **lock** block in their backend configuration and configure the desired locking mechanism. For example, to enable DynamoDB-based state locking with an Amazon S3 backend, users can configure the backend like this:

hclCopy code

```
terraform { backend "s3" { bucket = "my-terraform-state-bucket" key = "terraform.tfstate" region = "us-west-2" encrypt = true dynamodb_table = "terraform-state-lock" lock { dynamodb_table = "terraform-state-lock" } } }
```

In this example, the **lock** block specifies that Terraform should use DynamoDB-based state locking with the specified DynamoDB table.

By implementing effective state management practices, organizations can improve the reliability, scalability, and collaboration of their infrastructure as code workflows with Terraform. Remote state management, state locking, and other state management features help ensure the consistency and integrity of infrastructure deployments, enabling organizations to manage infrastructure at scale with confidence. With proper state management, organizations can automate infrastructure provisioning, track changes, and enforce governance and compliance requirements effectively.

Using remote backends for state storage is a fundamental practice in Terraform for managing infrastructure as code (IaC) workflows effectively. Remote backends provide a centralized and shared location to store Terraform state files, allowing teams to collaborate on infrastructure configurations seamlessly and ensuring consistency and reliability across environments. Terraform supports various types of remote backends, including cloud object storage services like Amazon S3, Azure Blob Storage, and Google Cloud Storage, as well as key-value stores like HashiCorp Consul and HashiCorp Terraform Cloud. Configuring a remote backend in Terraform involves specifying the backend type and configuration settings, either directly in the Terraform configuration file or via CLI commands. One of the most commonly used remote backends for Terraform is Amazon S3, a highly scalable and durable object storage service provided by Amazon Web Services (AWS). Configuring Terraform to use Amazon S3 as a remote backend involves creating an S3 bucket to store the Terraform state file and configuring the Terraform backend settings to point to the bucket.

To configure Terraform to use Amazon S3 as a remote backend, users first need to create an S3 bucket using the AWS Management Console or the AWS CLI. For example, to create an S3 bucket named **terraform-state-bucket** in the **us-west-2** region, users can run the following AWS CLI command:

bashCopy code

```
aws s3 mb s3://terraform-state-bucket --region us-west-2
```

Once the S3 bucket is created, users can configure the Terraform backend settings in their Terraform configuration file (**backend.tf**) to specify the S3 bucket as the remote backend.

hclCopy code

```
terraform { backend "s3" { bucket = "terraform-state-bucket" key = "terraform.tfstate" region = "us-west-2" encrypt = true dynamodb_table = "terraform-state-lock" } }
```

In this example, the **s3** backend type is specified, along with the S3 bucket name, key prefix, AWS region, encryption settings, and an optional DynamoDB table for state locking.

Additionally, users can configure encryption settings for the remote state data to ensure data security and compliance requirements. Terraform supports server-side encryption (SSE) for S3 buckets, allowing users to encrypt their state data using AWS-managed encryption keys (SSE-S3) or customer-managed encryption keys (SSE-KMS).

To enable SSE for an S3 bucket, users can specify the encryption settings when creating the bucket or update the bucket configuration using the AWS Management Console or the AWS CLI. For example, to enable SSE-S3 encryption for the **terraform-state-bucket**, users can run the following AWS CLI command:

bashCopy code

```
aws s3api put-bucket-encryption --bucket terraform-state-bucket --server-side-encryption-configuration '{"Rules": [{"ApplyServerSideEncryptionByDefault": {"SSEAlgorithm": "AES256"}}]}'
```

In this example, SSE-S3 encryption is enabled for the **terraform-state-bucket**, ensuring that the state data stored in the bucket is encrypted at rest using AES-256 encryption.

Moreover, Terraform supports state locking mechanisms to prevent concurrent modifications to the Terraform state by multiple users or processes. State locking helps prevent race conditions and conflicts that can arise when multiple users attempt to apply changes to the same infrastructure simultaneously. Terraform supports state locking using various mechanisms, including file-based locks, Consul locks, and DynamoDB locks, depending on the chosen backend.

To enable state locking with an S3 backend, users can specify the **lock** block in their backend configuration and configure the desired locking mechanism. For example, to enable DynamoDB-based state locking with the **terraform-state-lock** table, users can configure the backend like this:

hclCopy code

```
terraform { backend "s3" { bucket = "terraform-state-
bucket" key = "terraform.tfstate" region = "us-west-2"
encrypt = true dynamodb_table = "terraform-state-
lock" lock { dynamodb_table = "terraform-state-lock" } }
}
```

In this example, the **lock** block specifies that Terraform should use DynamoDB-based state locking with the specified DynamoDB table.

By using remote backends for state storage in Terraform, organizations can centralize and secure their infrastructure state data, enable collaboration and consistency across teams, and enhance the reliability and scalability of their infrastructure as code workflows. Remote backends provide a robust and scalable solution for managing Terraform state data, allowing organizations to deploy and manage infrastructure with confidence in dynamic and distributed environments.

Chapter 8: Terraform Best Practices and Conventions

Terraform coding standards are essential guidelines and best practices that help ensure consistency, readability, maintainability, and reliability of Terraform configurations. Adhering to coding standards fosters collaboration among team members, improves code quality, and reduces errors and inconsistencies in infrastructure deployments. Terraform coding standards cover various aspects of Terraform configuration files, including file structure, naming conventions, formatting, documentation, and usage of Terraform features and constructs. By following coding standards, organizations can streamline their Terraform development workflows, enhance code reviews, and build more robust and scalable infrastructure deployments.

One fundamental aspect of Terraform coding standards is file structure and organization. It is essential to maintain a consistent directory structure for Terraform projects to make it easier for team members to navigate and understand the project layout. A typical Terraform project may include directories for storing Terraform configuration files, module definitions, variables, outputs, and other supporting files. For example, a common directory structure for a Terraform project might look like this: cssCopy code

```
terraform-project/ ├── main.tf ├── variables.tf ├──
outputs.tf ├── modules/ | ├── module1/ | | ├──
main.tf | | ├── variables.tf | | └── outputs.tf | └──
module2/ | ├── main.tf | ├── variables.tf | └──
outputs.tf └── README.md
```

In this example, the **main.tf**, **variables.tf**, and **outputs.tf** files contain the main Terraform configuration, variable definitions, and output definitions, respectively. The **modules** directory contains subdirectories for modularizing Terraform configurations into reusable modules, each with its own **main.tf**, **variables.tf**, and **outputs.tf** files. Additionally, the project may include a **README.md** file for documentation purposes.

Another important aspect of Terraform coding standards is naming conventions for resources, variables, outputs, and other entities in Terraform configurations. Consistent and descriptive naming conventions help improve code readability and understandability, making it easier for team members to identify and use resources and variables. It is recommended to use descriptive names that reflect the purpose and function of each resource or variable. For example, instead of using generic names like **instance** or **subnet**, consider using more descriptive names like **web_server_instance** or **public_subnet**.

In addition to naming conventions, Terraform coding standards also include guidelines for formatting

Terraform configuration files to ensure consistency and readability. Terraform configurations are typically written in HashiCorp Configuration Language (HCL), which has its own syntax and formatting conventions. It is essential to follow consistent indentation, spacing, and alignment practices to make the code easier to read and understand. Tools like **terraform fmt** can automatically format Terraform configuration files according to the specified style guidelines. For example, to format all Terraform configuration files in a directory, users can run the following command:
bashCopy code

```
terraform fmt
```

This command formats all **.tf** files in the current directory and its subdirectories according to the default formatting rules.

Furthermore, Terraform coding standards may include guidelines for documenting Terraform configurations to provide context, usage instructions, and other relevant information for team members. Documentation can be added directly within the Terraform configuration files using comments or in separate documentation files. It is essential to document the purpose, usage, inputs, outputs, and any other relevant details for each Terraform resource, module, variable, and output.

Additionally, Terraform coding standards may include guidelines for leveraging Terraform features and constructs effectively and efficiently. This includes using Terraform modules to modularize and reuse infrastructure configurations, leveraging Terraform's built-in functions and operators, and adopting best practices for managing state, variables, outputs, and dependencies. By following these guidelines, organizations can maximize the benefits of Terraform and build scalable, maintainable, and reliable infrastructure deployments.

In summary, Terraform coding standards are essential guidelines and best practices that help ensure consistency, readability, maintainability, and reliability of Terraform configurations. By adhering to coding standards, organizations can streamline their Terraform development workflows, improve code quality, and build more robust and scalable infrastructure deployments. Terraform coding standards cover various aspects of Terraform configurations, including file structure, naming conventions, formatting, documentation, and usage of Terraform features and constructs, providing a comprehensive framework for building and managing infrastructure as code.

Infrastructure as code (IaC) best practices are essential guidelines and principles that help organizations effectively manage and automate their infrastructure deployments using code. These best practices encompass various aspects of IaC, including code organization, version control, automation, testing, security, and collaboration. By following IaC best practices, organizations can improve the reliability, scalability, security, and efficiency of their infrastructure deployments, streamline development workflows, and enhance collaboration among teams.

One of the foundational principles of IaC best practices is treating infrastructure configurations as code and applying software engineering practices to manage them effectively. This includes using version control systems like Git to track changes to infrastructure code, enabling collaboration, code reviews, and rollbacks. Git provides powerful version control capabilities, allowing users to manage changes, branches, and releases effectively. Organizations should adopt a version control workflow that suits their development processes, such as GitFlow or GitHub Flow, and establish branching, tagging, and merging conventions to manage infrastructure code changes.

To initialize a Git repository for an infrastructure project, users can navigate to the project directory and run the following command:

bashCopy code

```
git init
```

This command initializes a new Git repository in the current directory, enabling version control for the infrastructure code.

Once the Git repository is initialized, users can add infrastructure configuration files to the repository, commit changes, create branches for feature development or bug fixes, and collaborate with team members using pull requests and code reviews. By leveraging Git and version control best practices, organizations can ensure traceability, accountability, and repeatability of infrastructure deployments.

Another important aspect of IaC best practices is automation, which involves automating the deployment, provisioning, and management of infrastructure using code and automation tools. Automation enables organizations to deploy and update infrastructure rapidly, consistently, and reliably, reducing manual errors and increasing productivity. Tools like Terraform, AWS CloudFormation, Ansible, and Puppet provide powerful automation capabilities for managing infrastructure as code.

For example, to automate the provisioning of infrastructure resources using Terraform, users can define infrastructure configurations in Terraform configuration files (.tf), initialize a Terraform workspace, and apply the configuration to create or update resources. The following commands demonstrate the basic workflow for using Terraform to provision infrastructure:

bashCopy code

terraform init

This command initializes the Terraform workspace by downloading provider plugins and modules specified in the configuration files.

bashCopy code

terraform plan

This command generates an execution plan showing the changes Terraform will make to the infrastructure.

bashCopy code

terraform apply

This command applies the execution plan and creates or updates infrastructure resources as specified in the configuration files.

By automating infrastructure provisioning and management tasks with Terraform and other automation tools, organizations can achieve faster time-to-market, reduce operational overhead, and improve infrastructure consistency and reliability.

Additionally, IaC best practices include incorporating testing into the infrastructure development lifecycle to validate configurations, ensure compliance, and prevent regressions. Automated testing helps identify errors and misconfigurations early in the development process, reducing the risk of deployment failures and security vulnerabilities. Organizations can use tools like Terratest, Kitchen-Terraform, and InSpec to automate infrastructure testing and validation.

To run automated tests for infrastructure configurations using Terratest, users can define test cases in Go code and execute them using the **go test** command. The following example demonstrates a basic test case for validating an AWS EC2 instance provisioned by Terraform:

goCopy code

```
package test import ( "testing"
"github.com/gruntwork-
io/terratest/modules/terraform"
"github.com/stretchr/testify/assert" ) func
TestTerraformEC2Instance(t *testing.T) { t.Parallel()
terraformOptions := &terraform.Options{
TerraformDir: "../examples/ec2-instance", } defer
terraform.Destroy(t, terraformOptions)
terraform.InitAndApply(t, terraformOptions)
instanceID := terraform.Output(t, terraformOptions,
"instance_id") assert.NotEmpty(t, instanceID) }
```

This test case validates that an EC2 instance provisioned by Terraform has a non-empty instance ID.

By incorporating automated testing into the infrastructure development process, organizations can ensure the reliability, correctness, and security of their infrastructure configurations.

Moreover, IaC best practices include implementing security controls and compliance measures to protect infrastructure resources, data, and applications. This includes following security best practices for access control, encryption, network security, identity management, and data protection. Organizations should leverage cloud-native security services and features provided by cloud providers like AWS, Azure, and Google Cloud Platform (GCP) to enforce security policies and monitor for security threats and vulnerabilities.

For example, to implement encryption for data at rest and in transit in AWS, users can leverage services like AWS Key Management Service (KMS), AWS Certificate Manager (ACM), and AWS CloudTrail. By encrypting sensitive data and communications, organizations can prevent unauthorized access and data breaches, ensuring the confidentiality and integrity of their infrastructure and applications.

In summary, infrastructure as code (IaC) best practices encompass various principles and techniques for managing and automating infrastructure deployments effectively. By treating infrastructure configurations as code, automating deployment tasks, incorporating testing into the development lifecycle, implementing security controls, and collaborating with team members using version control and automation tools, organizations can build and manage infrastructure more reliably, securely, and efficiently. IaC best practices enable organizations to achieve greater agility, scalability, and resilience in their infrastructure deployments, enabling them to adapt to changing business requirements and technology trends effectively.

Chapter 9: Deploying and Managing Applications on AWS with Terraform

Deploying applications with Terraform is a crucial aspect of infrastructure as code (IaC) workflows, enabling organizations to automate the provisioning and management of infrastructure resources required to host and run applications. Terraform provides a declarative approach to defining infrastructure configurations, allowing users to describe the desired state of the infrastructure using Terraform configuration files and apply those configurations to create, update, or destroy resources. Deploying applications with Terraform involves defining infrastructure components such as virtual machines, containers, databases, networking configurations, and other necessary resources, as well as specifying how those resources are interconnected and configured to support the application.

One of the primary benefits of deploying applications with Terraform is the ability to manage infrastructure and application deployments together as code, providing a unified and repeatable approach to provisioning and managing both infrastructure and application components. By defining infrastructure configurations in Terraform configuration files, organizations can version control, test, and automate the deployment process, ensuring consistency and reliability across environments.

To deploy applications with Terraform, users typically start by defining the infrastructure components required to host the application in Terraform configuration files. This may include defining compute instances, storage resources, networking configurations, security groups, load balancers, and other necessary infrastructure components. Users can use Terraform's declarative language, HashiCorp Configuration Language (HCL), to describe the desired state of the infrastructure and specify the relationships and dependencies between resources.

Once the infrastructure configurations are defined, users can initialize a Terraform workspace and apply the configurations to create or update the infrastructure resources. The following commands demonstrate the basic workflow for deploying applications with Terraform:

bashCopy code

```
terraform init
```

This command initializes the Terraform workspace by downloading provider plugins and modules specified in the configuration files.

bashCopy code

```
terraform plan
```

This command generates an execution plan showing the changes Terraform will make to the infrastructure based on the defined configurations.

bashCopy code

terraform apply

This command applies the execution plan and creates or updates infrastructure resources as specified in the configuration files.

By following this workflow, users can deploy applications with Terraform in a controlled and repeatable manner, ensuring that the infrastructure is provisioned and configured correctly to support the application's requirements.

In addition to defining infrastructure resources, deploying applications with Terraform often involves configuring application-specific settings and dependencies, such as environment variables, application configuration files, database connections, and integrations with other services. Terraform allows users to parameterize configurations using variables and pass values dynamically to resources during deployment, enabling flexible and customizable deployments.

For example, users can define variables for application settings in Terraform configuration files and pass values to those variables using Terraform CLI flags or input variables files. The following example demonstrates how to define and use variables for configuring an application deployment with Terraform:

```
hclCopy code
# Define variables for application settings variable "app_port" { description = "The port on which the
```

application listens" type = number default = 8080 } variable "db_host" { description = "The hostname of the database server" type = string } # Configure application deployment resource "aws_instance" "app_server" { # Instance settings ami = "ami-12345678" instance_type = "t2.micro" # Application configuration tags = { Name = "app-server" } # Application-specific settings user_data = <<-EOF #!/bin/bash echo "Application Port: ${var.app_port}" echo "Database Host: ${var.db_host}" # Install and configure application... EOF }

In this example, two variables (**app_port** and **db_host**) are defined to specify application-specific settings such as the application port and the hostname of the database server. These variables can be passed values dynamically during deployment using Terraform CLI flags or input variables files, allowing users to customize the application deployment for different environments or configurations.

Furthermore, deploying applications with Terraform often involves managing dependencies between infrastructure resources and coordinating the deployment process with other tools and services. Terraform provides various features and constructs for managing dependencies and orchestrating complex deployment workflows, such as resource dependencies, provisioners, and external data sources.

For example, users can define dependencies between infrastructure resources in Terraform configuration files to ensure that resources are created or updated in the correct order. Additionally, Terraform allows users to define provisioners to execute scripts or commands on provisioned resources during deployment, enabling tasks such as software installation, configuration, and application deployment.

Overall, deploying applications with Terraform offers organizations a powerful and flexible approach to provisioning and managing infrastructure resources required to host and run applications. By defining infrastructure configurations as code and automating the deployment process with Terraform, organizations can achieve consistent, reliable, and scalable deployments, streamline development workflows, and accelerate time-to-market for their applications.

Managing the application lifecycle with Terraform involves orchestrating the deployment, scaling, updating, and retiring of applications and their associated infrastructure using Terraform's infrastructure as code (IaC) capabilities. Terraform provides a declarative approach to defining and managing infrastructure configurations, allowing organizations to automate the entire lifecycle of applications from provisioning to decommissioning. By treating infrastructure and application configurations as code, organizations can achieve consistency, repeatability, and scalability in managing application deployments across different environments.

The first step in managing the application lifecycle with Terraform is defining the infrastructure resources and configurations required to support the application. This involves specifying compute instances, storage resources, networking configurations, security settings, and other infrastructure components in Terraform configuration files. Organizations can use Terraform modules to modularize and reuse infrastructure configurations, enabling them to define common patterns and best practices for deploying applications.

To define infrastructure configurations for an application deployment, users can create Terraform configuration files (**.tf**) and specify the desired state of the infrastructure using Terraform's declarative language, HashiCorp Configuration Language (HCL). For example, to define an AWS EC2 instance for hosting a web application, users can create a Terraform configuration file (**main.tf**) with the following contents:

hclCopy code

main.tf provider "aws" { region = "us-east-1" } resource "aws_instance" "web_server" { ami = "ami-

12345678" instance_type = "t2.micro" tags = { Name = "web-server" } }

In this example, the **aws_instance** resource defines an EC2 instance with the specified AMI (Amazon Machine Image) and instance type. The **tags** attribute is used to assign metadata to the instance, such as its name.

Once the infrastructure configurations are defined, users can initialize a Terraform workspace and apply the configurations to provision the infrastructure resources. The following commands demonstrate the basic workflow for managing the application lifecycle with Terraform:

bashCopy code

```
terraform init
```

This command initializes the Terraform workspace by downloading provider plugins and modules specified in the configuration files.

bashCopy code

```
terraform plan
```

This command generates an execution plan showing the changes Terraform will make to the infrastructure based on the defined configurations.

bashCopy code

```
terraform apply
```

This command applies the execution plan and creates or updates infrastructure resources as specified in the configuration files.

By following this workflow, users can provision the infrastructure resources required to support the application deployment in a repeatable and automated manner.

In addition to provisioning infrastructure resources, managing the application lifecycle with Terraform also involves configuring and deploying the application code and dependencies. This may include tasks such as deploying application artifacts, configuring environment variables, setting up databases, configuring networking, and integrating with other services.

Terraform provides various mechanisms for managing application configurations and dependencies, including user data scripts, provisioners, and external data sources. For example, users can use Terraform's **user_data** attribute to pass initialization scripts or configuration commands to provisioned instances. Additionally, Terraform's provisioners allow users to execute scripts or commands on provisioned resources during deployment, enabling tasks such as software installation, configuration, and application deployment.

To illustrate, consider the following Terraform configuration file that defines an EC2 instance with a user data script to deploy a web application:

hclCopy code

```
# main.tf provider "aws" { region = "us-east-1" }
resource "aws_instance" "web_server" { ami = "ami-
```

12345678" instance_type = "t2.micro" tags = { Name = "web-server" } user_data = <<-EOF #!/bin/bash sudo yum update -y sudo yum install -y httpd sudo systemctl start httpd sudo systemctl enable httpd echo "<html><h1>Hello, World!</h1></html>" > /var/www/html/index.html EOF }

In this example, the **user_data** attribute contains a bash script that updates the package repositories, installs the Apache web server, starts the web server, enables it to start on boot, and creates a simple HTML file to serve as the application.

Moreover, managing the application lifecycle with Terraform involves scaling the infrastructure resources to accommodate changes in demand or workload requirements. Terraform allows users to define scaling policies and configurations to automatically adjust the number of compute instances, storage resources, and other infrastructure components based on predefined conditions or metrics. For example, users can use Terraform's **autoscaling** feature to define auto scaling groups that automatically add or remove instances based on CPU utilization, network traffic, or other metrics.

To demonstrate, consider the following Terraform configuration file that defines an auto scaling group for a web application:

hclCopy code

main.tf provider "aws" { region = "us-east-1" } resource "aws_launch_configuration"

```
"web_server_config" { image_id = "ami-12345678"
instance_type = "t2.micro" } resource
"aws_autoscaling_group" "web_server_group" {
launch_configuration =
aws_launch_configuration.web_server_config.id
min_size = 2 max_size = 10 desired_capacity = 5 }
```

In this example, the **aws_launch_configuration** resource defines the configuration for the instances launched by the auto scaling group, including the AMI and instance type. The **aws_autoscaling_group** resource defines the auto scaling group with a minimum size of 2 instances, a maximum size of 10 instances, and a desired capacity of 5 instances.

By defining auto scaling policies and configurations in Terraform, organizations can ensure that their applications can dynamically scale to handle changes in demand or workload without manual intervention.

Furthermore, managing the application lifecycle with Terraform involves updating and modifying the infrastructure configurations and application deployments as requirements evolve over time. Terraform allows users to make changes to infrastructure configurations and apply those changes to update or modify existing resources. Organizations can use Terraform's plan and apply workflow to review and apply changes safely and predictably, ensuring that updates are applied consistently across environments.

To update existing infrastructure resources with Terraform, users can make changes to the Terraform configuration files and run the following commands:
bashCopy code

terraform plan

This command generates an execution plan showing the changes Terraform will make to the infrastructure based on the updated configurations.
bashCopy code

terraform apply

This command applies the execution plan and updates or modifies existing infrastructure resources as specified in the updated configuration files.

By following this workflow, users can apply changes to infrastructure configurations and application deployments in a controlled and automated manner, reducing the risk of errors and ensuring consistency across environments.

Moreover, managing the application lifecycle with Terraform includes monitoring, logging, and troubleshooting applications and infrastructure deployments to ensure their reliability and performance. Terraform integrates with various monitoring and logging services provided by cloud providers, allowing organizations to collect metrics, monitor performance, and troubleshoot issues in real-time.

For example, users can use Terraform to define configurations for setting up monitoring and logging services such as AWS CloudWatch, Azure Monitor, or

Google Cloud Monitoring. By defining monitoring and logging configurations as code, organizations can automate the setup and configuration of these services and ensure that they are consistently applied across environments.

Additionally, Terraform provides features for managing infrastructure drift and enforcing desired state configurations. Terraform's state management capabilities allow users to track the current state of infrastructure resources and detect any deviations from the desired state defined in the Terraform configuration files. Organizations can use Terraform's state commands to inspect, import, and manage the state of infrastructure resources, ensuring that they remain in sync with the desired configurations.

To summarize, managing the application lifecycle with Terraform involves defining and provisioning infrastructure resources, deploying and configuring applications, scaling infrastructure to accommodate changes in demand, updating and modifying configurations, monitoring and troubleshooting deployments, and enforcing desired state configurations. By leveraging Terraform's infrastructure as code capabilities, organizations can automate and streamline the entire lifecycle of their applications, improving agility, reliability, and scalability in managing application deployments across different environments.

Chapter 10: Troubleshooting and Debugging Terraform Deployments

Diagnosing Terraform errors is a critical aspect of managing infrastructure as code (IaC) workflows, as it allows users to identify and resolve issues that may arise during the provisioning and management of infrastructure resources. Terraform provides various mechanisms and tools for diagnosing errors, including error messages, logs, debugging features, and community resources. By understanding common types of errors, interpreting error messages, and using debugging techniques, users can effectively diagnose and troubleshoot Terraform errors to ensure the reliability and stability of their infrastructure deployments.

When encountering errors in Terraform, the first step is to understand the nature of the error and identify its root cause. Terraform provides detailed error messages that describe the nature of the error, its location in the configuration files, and potential solutions. Users can use the **terraform plan** and **terraform apply** commands to generate execution plans and apply changes to infrastructure resources, respectively. If there are errors in the configuration files, Terraform will display error messages indicating the cause of the error and suggestions for resolution.

For example, if there is a syntax error in a Terraform configuration file, Terraform will display an error message indicating the location of the error and the specific syntax issue. Users can inspect the error message to identify the syntax error and correct it accordingly. The following command demonstrates how to use **terraform plan** to generate an execution plan and identify syntax errors:

bashCopy code

```
terraform plan
```

This command generates an execution plan based on the Terraform configuration files and displays any errors or warnings encountered during the planning process. Users can review the error messages to identify syntax errors and make necessary corrections to the configuration files.

In addition to error messages, Terraform generates logs that provide information about the execution of Terraform commands, including details about resource creation, modification, and deletion. Users can inspect the logs to identify any issues or unexpected behavior during the execution of Terraform commands. Terraform logs are stored in log files located in the **.terraform** directory within the Terraform working directory.

To view Terraform logs, users can use a text editor or command-line tools to open the log files and inspect the contents. For example, users can use the **cat** command to display the contents of a log file in the terminal:

bashCopy code

```
cat .terraform/terraform.log
```

This command displays the contents of the Terraform log file, allowing users to inspect the log entries and identify any errors or issues encountered during the execution of Terraform commands.

In addition to error messages and logs, Terraform provides debugging features that allow users to enable verbose logging and trace execution to diagnose errors and troubleshoot issues. Users can enable debug mode by setting the **TF_LOG** environment variable to **DEBUG** before executing Terraform commands. This enables verbose logging, which provides detailed information about the execution of Terraform commands, including HTTP requests, API responses, and internal operations.

To enable debug mode in Terraform, users can set the **TF_LOG** environment variable to **DEBUG** using the following command:

bashCopy code

```
export TF_LOG=DEBUG
```

This command sets the **TF_LOG** environment variable to **DEBUG**, enabling debug mode for Terraform commands executed in the current terminal session. Users can then execute Terraform commands as usual, and Terraform will output debug log messages to the terminal, providing detailed information about the execution of the commands.

Furthermore, when diagnosing Terraform errors, users can leverage community resources, including forums, discussion groups, and documentation, to seek assistance from other Terraform users and experts. The Terraform community is active and supportive, with many users willing to provide help and guidance on diagnosing and troubleshooting Terraform errors. Users can search for error messages or specific issues in online forums and discussion groups, such as the Terraform GitHub repository, HashiCorp Community Forum, and Stack Overflow, to find solutions to common problems and receive advice from experienced users.

Additionally, Terraform documentation provides comprehensive information and guidance on troubleshooting common errors, debugging techniques, best practices, and recommended approaches for diagnosing and resolving issues. Users can consult the Terraform documentation to learn more about specific error messages, debugging features, and troubleshooting strategies, as well as access tutorials, guides, and examples to help them troubleshoot Terraform errors effectively.

In summary, diagnosing Terraform errors involves understanding error messages, inspecting logs, enabling debug mode, and leveraging community resources and documentation to identify and resolve issues. By using error messages, logs, debugging features, and community resources effectively, users can diagnose and troubleshoot Terraform errors efficiently, ensuring the reliability and stability of their infrastructure deployments.

Debugging Terraform deployments is an essential skill for infrastructure as code (IaC) practitioners, enabling them to identify and resolve issues that may arise during the provisioning and management of infrastructure resources. Terraform deployments can encounter various types of errors and unexpected behaviors, ranging from syntax errors in configuration files to issues with resource dependencies and provider configurations. To effectively debug Terraform deployments, users can leverage a combination of error messages, logs, debugging features, and troubleshooting techniques to diagnose and address issues as they arise.

When debugging Terraform deployments, the first step is to understand the nature of the issue and identify its root cause. Terraform provides detailed error messages that describe the nature of the error, its location in the configuration files, and potential solutions. Users can use the **terraform plan** and **terraform apply** commands to generate execution plans and apply changes to infrastructure resources, respectively. If there are errors in the configuration files or issues with resource dependencies, Terraform will display error messages indicating the cause of the error and suggestions for resolution.

For example, if there is a syntax error in a Terraform configuration file, Terraform will display an error message indicating the location of the error and the specific syntax issue. Users can inspect the error message to identify the syntax error and correct it accordingly. The following command demonstrates how to use **terraform plan** to generate an execution plan and identify syntax errors:

bashCopy code

```
terraform plan
```

This command generates an execution plan based on the Terraform configuration files and displays any errors or warnings encountered during the planning process. Users can review the error messages to identify syntax errors and make necessary corrections to the configuration files.

In addition to error messages, Terraform generates logs that provide information about the execution of Terraform commands, including details about resource creation, modification, and deletion. Users can inspect the logs to identify any issues or unexpected behavior during the execution of Terraform commands. Terraform logs are stored in log files located in the **.terraform** directory within the Terraform working directory.

To view Terraform logs, users can use a text editor or command-line tools to open the log files and inspect the contents. For example, users can use the **cat** command to display the contents of a log file in the terminal:

bashCopy code

```
cat .terraform/terraform.log
```

This command displays the contents of the Terraform log file, allowing users to inspect the log entries and identify any errors or issues encountered during the execution of Terraform commands.

In addition to error messages and logs, Terraform provides debugging features that allow users to enable verbose logging and trace execution to diagnose errors and troubleshoot issues. Users can enable debug mode by setting the **TF_LOG** environment variable to **DEBUG** before executing Terraform commands. This enables verbose logging, which provides detailed information about the execution of Terraform commands, including HTTP requests, API responses, and internal operations.

To enable debug mode in Terraform, users can set the **TF_LOG** environment variable to **DEBUG** using the following command:

bashCopy code

```
export TF_LOG=DEBUG
```

This command sets the **TF_LOG** environment variable to **DEBUG**, enabling debug mode for Terraform commands executed in the current terminal session. Users can then execute Terraform commands as usual, and Terraform will output debug log messages to the terminal, providing detailed information about the execution of the commands.

Furthermore, when debugging Terraform deployments, users can leverage community resources, including forums, discussion groups, and documentation, to seek assistance from other

Terraform users and experts. The Terraform community is active and supportive, with many users willing to provide help and guidance on debugging and troubleshooting Terraform deployments. Users can search for error messages or specific issues in online forums and discussion groups, such as the Terraform GitHub repository, HashiCorp Community Forum, and Stack Overflow, to find solutions to common problems and receive advice from experienced users.

Additionally, Terraform documentation provides comprehensive information and guidance on debugging techniques, troubleshooting strategies, and best practices for diagnosing and resolving issues. Users can consult the Terraform documentation to learn more about specific error messages, debugging features, and troubleshooting approaches, as well as access tutorials, guides, and examples to help them debug Terraform deployments effectively.

In summary, debugging Terraform deployments involves understanding error messages, inspecting logs, enabling debug mode, and leveraging community resources and documentation to identify and resolve issues. By using error messages, logs, debugging features, and community resources effectively, users can diagnose and troubleshoot Terraform deployments efficiently, ensuring the reliability and stability of their infrastructure deployments.

BOOK 2
MASTERING TERRAFORM
ADVANCED TECHNIQUES FOR AWS CLOUD
AUTOMATION

ROB BOTWRIGHT

Chapter 1: Advanced Terraform Configuration Patterns

Dynamic configuration using variables is a fundamental concept in Terraform that allows users to parameterize and customize their infrastructure configurations based on varying requirements and conditions. By defining variables in Terraform configuration files, users can create reusable templates and templates that adapt to different environments, workloads, and use cases. Terraform variables can be used to specify values for resource attributes, module inputs, and provider configurations, enabling users to create flexible and dynamic infrastructure deployments that can be easily customized and scaled as needed.

To define variables in Terraform, users can create variable declarations in Terraform configuration files using the **variable** block. For example, to define a variable named **instance_type** to specify the instance type for an AWS EC2 instance, users can add the following variable declaration to their Terraform configuration file:

hclCopy code

variable "instance_type" { type = string description = "The instance type for the EC2 instance" default = "t2.micro" }

In this example, the **variable** block defines a variable named **instance_type** with a string type, a description, and a default value of **"t2.micro"**. Users can customize the value of the **instance_type** variable by providing a different value when running Terraform commands or by specifying a value in a variable file.

To provide a value for a variable when running Terraform commands, users can use the **-var** flag followed by the variable name and value. For example, to specify a different instance type for the **instance_type** variable, users can run the following Terraform command:

bashCopy code

terraform plan -var="instance_type=t3.medium"

This command specifies the value **"t3.medium"** for the **instance_type** variable when generating an execution plan.

Alternatively, users can create variable files to specify values for variables and provide the variable file to Terraform commands using the **-var-file** flag. Variable files are JSON or HCL files that contain key-value pairs representing variable names and values. For example, users can create a variable file named **variables.tfvars** with the following contents:

hclCopy code

instance_type = "t3.medium"

To use the variable file when running Terraform commands, users can use the following command:

bashCopy code

terraform plan -var-file="variables.tfvars"

This command specifies the values defined in the **variables.tfvars** file for variables when generating an execution plan.

In addition to specifying values for variables, users can use Terraform's interpolation syntax to reference variables and use their values in resource configurations. Interpolation syntax allows users to embed variables and expressions within strings and configurations, enabling dynamic and flexible resource definitions.

For example, users can reference the **instance_type** variable in an AWS EC2 instance resource definition as follows:

hclCopy code

```
resource "aws_instance" "web_server" { ami = "ami-12345678" instance_type = var.instance_type tags = { Name = "web-server" } }
```

In this example, the **instance_type** attribute of the **aws_instance** resource is set to the value of the **instance_type** variable using Terraform's interpolation syntax (**var.instance_type**). This allows users to dynamically specify the instance type for the EC2 instance based on the value of the **instance_type** variable.

Furthermore, Terraform supports various types of variables, including string, number, bool, list, and map types, allowing users to define variables for different types of values and data structures. Users can also specify validation rules, constraints, and defaults for variables to enforce data integrity and provide default values for optional variables.

To define a variable with a specific type, users can use the **type** attribute in the **variable** block. For example, to define a variable named **subnet_ids** to specify a list of subnet IDs for an AWS VPC resource, users can use the following variable declaration:

hclCopy code

variable "subnet_ids" { type = list(string) description = "A list of subnet IDs for the VPC" default = ["subnet-12345678", "subnet-23456789"] }

In this example, the **subnet_ids** variable is defined as a list of strings (**list(string)**), allowing users to specify multiple subnet IDs for the VPC. The variable has a default value of **["subnet-12345678", "subnet-23456789"]**, which users can override when running Terraform commands.

Overall, dynamic configuration using variables is a powerful feature in Terraform that enables users to create flexible, reusable, and customizable infrastructure configurations. By defining variables, referencing them in resource configurations, and specifying values using interpolation syntax, users can create dynamic infrastructure deployments that adapt to different environments and use cases. Additionally, Terraform's support for various types of variables and validation rules allows users to enforce data integrity and provide default values for variables, ensuring consistent and reliable infrastructure deployments.

Terraform template rendering techniques are crucial for dynamically generating configuration files, scripts, and other artifacts based on template files and variable inputs. Template rendering enables users to create reusable templates that adapt to various environments, configurations, and use cases. Terraform provides built-in template rendering capabilities using the **templatefile** function, allowing users to generate dynamic content based on input variables and template files. Additionally, users can leverage external template rendering tools and libraries, such as **consul-template** or **Jinja2**, to create more complex and flexible templates for their Terraform configurations.

The **templatefile** function in Terraform allows users to render templates using a combination of variables and template files. This function takes a template file as input and interpolates variables and expressions within the template to generate dynamic content. Users can define variables and values in their Terraform configuration files and reference them in the template using Terraform's interpolation syntax. The **templatefile** function then processes the template file, replacing variable references with their corresponding values, and generates the rendered output.

To use the **templatefile** function in Terraform, users can create a template file with placeholders for variables and expressions and use the **templatefile** function to render the template. For example, users can create a template file named **app_config.tpl** with the following content:

hclCopy code

```
# App Configuration app_name = "${app_name}"
environment = "${environment}" region = "${region}"
instance_type = "${instance_type}"
```

In this template file, **${}** placeholders represent variables that will be replaced with their corresponding values during template rendering. Users can then use the **templatefile** function in their Terraform configuration files to render the template and generate the rendered output. The following example demonstrates how to use the **templatefile** function to render the template file and generate the rendered output:

hclCopy code

```
data "template_file" "app_config" { template =
file("app_config.tpl") vars = { app_name = "my-app"
environment = "production" region = "us-west-2"
instance_type = "t2.micro" } } output
"rendered_app_config" { value =
data.template_file.app_config.rendered }
```

In this example, the **data "template_file"** block defines a template file data source named **app_config** and specifies the template file (**app_config.tpl**) and variable values (**app_name, environment, region, instance_type**). The **templatefile** function renders the template file using the specified variable values, and the rendered output is stored in the **rendered_app_config** output variable.

Alternatively, users can leverage external template rendering tools and libraries, such as **consul-template** or **Jinja2**, to create more complex and flexible templates for their Terraform configurations. These tools provide advanced templating features, such as conditionals, loops, and functions, allowing users to create dynamic and expressive templates that meet their specific requirements.

For example, **consul-template** is a popular template rendering tool that integrates with Consul and other data sources to dynamically generate configuration files based on template files and key-value data. Users can define templates using the Go template language and reference data from Consul, environment variables, or other sources within the templates. **consul-template** then watches for changes to the data sources and automatically updates the rendered output whenever the data changes.

To use **consul-template** with Terraform, users can create a template file using the Go template language and use **consul-template** to render the template and generate the rendered output. For example, users can create a template file named **app_config.ctmpl** with the following content:

goCopy code

```
# App Configuration app_name = {{ key "app/name" }} environment = {{ env "ENVIRONMENT" }} region = {{ key "aws/region" }} instance_type = {{ key "instance/type" }}
```

In this template file, **{{}}** placeholders represent expressions that will be replaced with their corresponding values during template rendering. Users can then use **consul-template** to render the template file and generate the rendered output. The following example demonstrates how to use **consul-template** to render the template file and generate the rendered output:

bashCopy code

```
consul-template                                     -
template="app_config.ctmpl:app_config.tf"
```

This command instructs **consul-template** to render the **app_config.ctmpl** template file and generate the rendered output in a file named **app_config.tf**. Users can then use the rendered output as a Terraform configuration file in their infrastructure deployments.

Overall, Terraform template rendering techniques enable users to create dynamic and flexible infrastructure configurations by generating configuration files, scripts, and other artifacts based on template files and variable inputs. By leveraging built-in template rendering capabilities and external template rendering tools and libraries, users can create reusable templates that adapt to different environments, configurations, and use cases, improving productivity and maintainability in their Terraform workflows.

Chapter 2: Terraform Providers and Plugins Deep Dive

Custom provider development in Terraform is a powerful technique that allows users to extend Terraform's capabilities by creating custom providers to interact with APIs, services, and resources that are not supported out-of-the-box. By developing custom providers, users can automate the management of infrastructure resources that are specific to their organization, industry, or use case, enabling them to leverage Terraform's infrastructure as code (IaC) approach for a wider range of applications and environments. Developing a custom provider involves several key steps, including defining resource types, implementing CRUD operations, handling authentication and authorization, and testing and debugging the provider code.

The first step in custom provider development is defining resource types and operations that the provider will support. Resource types represent the different types of infrastructure resources that users can manage using the custom provider, such as virtual machines, databases, or networking components. Users define resource types by creating schema files that specify the attributes and properties of each resource type, including their names, types, descriptions, and constraints. Additionally, users

95

define operations for each resource type, such as create, read, update, and delete (CRUD) operations, to enable users to interact with and manage the resources.

To define resource types and operations, users create schema files in the Terraform provider configuration format (.tf files) and specify the resource attributes, properties, and operations using the Terraform schema language. For example, users can define a resource type for managing virtual machines in a cloud provider by creating a schema file named **virtual_machine.tf** with the following content:

hclCopy code

```
resource "mycloud_virtual_machine" "instance" {
name = string size = string image = string region = string }
```

In this schema file, the **mycloud_virtual_machine** resource type is defined with attributes for **name**, **size**, **image**, and **region**. Users can define additional attributes and properties as needed to represent the desired configuration for virtual machines in the custom provider.

Once resource types and operations are defined, users implement CRUD operations for each resource type to enable users to create, read, update, and delete resources using the custom provider. CRUD operations are implemented as functions or methods in the provider code that interact with the underlying APIs, services, or resources to perform the requested operations. Users implement CRUD operations by writing code in a programming language supported by Terraform providers, such as Go, to handle the necessary API requests, responses, and error handling logic.

To implement CRUD operations, users create provider packages in Go and define functions or methods for each CRUD operation for each resource type. For example, users can define functions for creating, reading, updating, and deleting virtual machines in the custom provider package as follows:

```go
goCopy code
func (c *Client) CreateVirtualMachine(ctx context.Context, vm *VirtualMachine) error { // Implement logic to create a virtual machine using the API } func (c *Client) ReadVirtualMachine(ctx context.Context, id string) (*VirtualMachine, error) { // Implement logic to read a virtual machine using the API } func (c *Client) UpdateVirtualMachine(ctx context.Context, id string, vm *VirtualMachine) error { // Implement logic to update a virtual machine using the API } func (c *Client) DeleteVirtualMachine(ctx context.Context, id string) error { // Implement logic to delete a virtual machine using the API }
```

In this example, functions are defined for creating, reading, updating, and deleting virtual machines in the custom provider package. Each function accepts input parameters, such as a context and resource identifier, and interacts with the underlying API to perform the requested operation.

Once CRUD operations are implemented, users handle authentication and authorization to ensure secure access to resources and prevent unauthorized access. Authentication and authorization mechanisms vary depending on the API, service, or resource being accessed and may involve API keys, OAuth tokens, or other authentication methods. Users implement authentication and authorization logic in the provider code to authenticate users, generate access tokens, and authorize API requests.

To handle authentication and authorization, users configure authentication settings in the provider configuration file and implement authentication logic in the provider code. For example, users can configure API keys or OAuth tokens as environment variables or configuration settings and use them to authenticate API requests in the provider code.

goCopy code

func (c *Client) Authenticate(ctx context.Context) error { // Implement logic to authenticate with the API using API keys or OAuth tokens }

In this example, the **Authenticate** function is defined to authenticate with the API using API keys or OAuth tokens. Users can implement logic to retrieve API keys or OAuth tokens from environment variables or configuration settings and use them to authenticate API requests.

Finally, users test and debug the custom provider code to ensure its functionality, reliability, and performance. Testing involves writing unit tests, integration tests, and end-to-end tests to validate the behavior of the provider code and ensure that it meets the specified requirements and expectations. Users use testing frameworks and tools, such as Go's testing package, to write and execute tests for the provider code and verify its correctness and robustness.

To test and debug the custom provider code, users create test cases and scenarios to cover different use cases, edge cases, and failure scenarios. Users execute tests locally or in a testing environment and analyze the test results to identify issues, errors, and performance bottlenecks. Users use debugging tools, logging, and error handling techniques to diagnose and resolve issues in the provider code and ensure its quality and stability.

Overall, custom provider development in Terraform is a complex and involved process that requires careful planning, design, implementation, testing, and debugging. By following best practices and techniques for defining resource types, implementing CRUD operations, handling authentication and authorization, and testing and debugging the provider code, users can create custom providers that extend Terraform's capabilities and enable automated management of infrastructure resources in a wide range of environments and use cases.

Plugin architecture and extensibility are fundamental concepts in the design and development of modern software systems, allowing for modularization, flexibility, and scalability. In the context of infrastructure as code (IaC) tools like Terraform, plugin architecture plays a crucial role in enabling users to extend the functionality of the core platform by developing custom plugins or providers. These plugins empower users to interact with a wide range of cloud providers, services, and resources beyond the capabilities provided by the default Terraform providers. Understanding plugin architecture and extensibility is essential for users who wish to customize and extend Terraform's functionality to meet their specific infrastructure requirements.

Terraform's plugin architecture follows a modular design approach, where each provider or plugin is encapsulated as a separate executable binary file that communicates with the Terraform core via a well-defined plugin interface. This interface allows Terraform to dynamically load and execute plugins at runtime, enabling users to interact with various cloud platforms and services seamlessly. By leveraging this plugin architecture, users can develop custom plugins or providers to integrate Terraform with proprietary or specialized systems, services, and APIs.

To develop custom providers for Terraform, users typically follow a set of best practices and guidelines provided by HashiCorp, the company behind Terraform. These guidelines include defining a provider schema, implementing CRUD (Create, Read, Update, Delete) operations for resource management, handling authentication and authorization, and supporting features such as import and state management. Additionally, users need to adhere to Terraform's plugin protocol, which specifies the communication protocol between Terraform and plugins, including data serialization formats, error handling, and lifecycle management.

The first step in developing a custom provider for Terraform is defining the provider schema, which describes the configuration options, resource types, and data sources supported by the provider. This schema is typically defined using HashiCorp Configuration Language (HCL) and includes information such as resource attributes, input variables, output values, and supported operations. Once the provider schema is defined, users can implement the CRUD operations for resource management, including resource creation, reading, updating, and deletion.

To implement CRUD operations for resource management, users need to define handlers or methods for each operation and integrate them into the provider's codebase. These handlers typically interact with the target infrastructure or service via

APIs, SDKs, or command-line tools, performing the necessary actions to create, read, update, or delete resources as requested by Terraform. Users also need to implement error handling, input validation, and state management to ensure the reliability and consistency of resource operations.

Authentication and authorization are critical aspects of custom provider development, as they determine how users authenticate and access resources managed by the provider. Depending on the target platform or service, users may need to implement various authentication mechanisms, such as API keys, OAuth tokens, or IAM roles, and handle authentication and authorization logic within the provider's codebase. Additionally, users need to support secure credential management and storage to ensure the confidentiality and integrity of sensitive information, such as access tokens or credentials.

Once the custom provider is developed and tested, users can distribute it as a standalone executable binary file or publish it to a repository or registry for broader community adoption. Terraform provides a mechanism for installing and managing third-party providers using the **terraform init** command, which initializes a Terraform configuration and installs any required providers or plugins specified in the configuration files. Users can specify the source URLs or paths of custom providers in their Terraform configurations, allowing Terraform to automatically download and install the providers during initialization.

For example, to install a custom provider named **myprovider**, users can use the following command:

bashCopy code

terraform init -plugin-dir=/path/to/custom/providers

This command initializes the Terraform configuration and instructs Terraform to search for custom providers in the specified directory (**/path/to/custom/providers**) and install them for use in the configuration.

Overall, plugin architecture and extensibility play a crucial role in enabling users to customize and extend Terraform's functionality to meet their specific infrastructure requirements. By understanding the principles of plugin development, users can develop custom providers, integrate Terraform with proprietary or specialized systems, and leverage the full power and flexibility of Terraform for infrastructure as code.

Chapter 3: Terraform State Management Strategies

State storage considerations are crucial when working with infrastructure as code (IaC) tools like Terraform, as they impact the reliability, scalability, and manageability of infrastructure deployments. In Terraform, state refers to a JSON file that contains information about the resources managed by Terraform, including resource metadata, dependencies, and attributes. This state file serves as the source of truth for Terraform's understanding of the infrastructure and is used to track changes, perform operations, and ensure consistency across deployments. Therefore, choosing an appropriate state storage backend is essential for maintaining the integrity and stability of Terraform workflows.

Terraform supports various state storage backends, each with its own advantages, limitations, and use cases. The default state storage backend in Terraform is the local backend, which stores the state file on the local filesystem of the machine running Terraform. While the local backend is convenient for testing and development purposes, it is not suitable for production environments due to its lack of durability, scalability, and concurrency support. Changes made to the infrastructure by different users or machines may result in conflicts or inconsistencies when using the local backend, making it unsuitable for collaborative or distributed workflows.

To address the limitations of the local backend, Terraform provides several remote state storage backends that offer improved reliability, scalability, and collaboration capabilities. One of the most commonly used remote state storage backends is the Amazon S3 backend, which stores the state file in an Amazon Simple Storage Service (S3) bucket. Using the S3 backend allows users to leverage S3's durability, availability, and versioning features, ensuring that the state file is securely stored and accessible from anywhere. Additionally, the S3 backend supports locking mechanisms to prevent concurrent state modifications, enabling safe and consistent Terraform operations in multi-user or multi-machine environments.

To configure Terraform to use the Amazon S3 backend for state storage, users can specify the backend configuration in their Terraform configuration files or using CLI commands. For example, users can create a **backend.tf** file with the following configuration:

hclCopy code

terraform { backend "s3" { bucket = "my-terraform-state" key = "terraform.tfstate" region = "us-west-2" dynamodb_table = "terraform-lock" } }

In this configuration, the **backend** block specifies the S3 backend configuration, including the S3 bucket name (**my-terraform-state**), the key prefix for the state file (**terraform.tfstate**), the AWS region (**us-west-2**), and an optional DynamoDB table name (**terraform-lock**) for state locking. Terraform automatically manages state locking when using the S3 backend, ensuring that concurrent Terraform operations are serialized and coordinated to prevent conflicts.

In addition to the Amazon S3 backend, Terraform supports other remote state storage backends, including Azure Blob Storage, Google Cloud Storage, and HashiCorp Consul. Each of these backends offers similar benefits in terms of reliability, scalability, and collaboration, allowing users to choose the backend that best fits their infrastructure environment and requirements. For example, users operating in a Microsoft Azure environment may prefer the Azure Blob Storage backend for its seamless integration with Azure services and ecosystem.

To configure Terraform to use the Azure Blob Storage backend for state storage, users can create a **backend.tf** file with the following configuration:

hclCopy code

```
terraform { backend "azurerm" { storage_account_name = "my-terraform-storage" container_name = "terraform-state" key = "terraform.tfstate" } }
```

In this configuration, the **backend** block specifies the Azure Blob Storage backend configuration, including the storage account name (**my-terraform-storage**), the container name (**terraform-state**), and the key prefix for the state file (**terraform.tfstate**). Terraform automatically manages state locking when using the Azure Blob Storage backend, ensuring safe and consistent Terraform operations in multi-user or multi-machine environments.

Overall, state storage considerations are critical for ensuring the reliability, scalability, and collaboration capabilities of Terraform workflows. By choosing an appropriate remote state storage backend and configuring Terraform accordingly, users can maintain the integrity and stability of their infrastructure deployments and streamline collaboration and coordination in distributed or team-based environments.

Advanced state locking mechanisms are essential for managing concurrent access to Terraform state files in collaborative or distributed environments, ensuring the integrity and consistency of infrastructure deployments. While Terraform's default state locking mechanism provides basic concurrency control using a single lock file, advanced state locking mechanisms offer more sophisticated capabilities, such as fine-grained locking, distributed locking, and integration with external systems. These mechanisms help prevent conflicts, data corruption, and race conditions when multiple users or processes interact with

Terraform concurrently, improving the reliability and scalability of Terraform workflows.

One of the most commonly used advanced state locking mechanisms is DynamoDB-based locking, which leverages Amazon DynamoDB to implement distributed locking for Terraform state files. DynamoDB is a fully managed NoSQL database service provided by Amazon Web Services (AWS) that offers fast and scalable performance, high availability, and built-in support for distributed locking and concurrency control. By using DynamoDB-based locking, users can ensure that Terraform operations are serialized and coordinated across multiple users or machines, preventing conflicts and data corruption. To configure Terraform to use DynamoDB-based locking for state locking, users need to create a DynamoDB table with the appropriate schema and permissions and configure Terraform to use this table for state locking. First, users can create a DynamoDB table named **terraform-lock** with a primary key named **LockID** using the AWS Management Console or CLI. For example, users can use the following AWS CLI command to create the DynamoDB table:

bashCopy code

```
aws dynamodb create-table \ --table-name
terraform-lock \ --attribute-definitions
AttributeName=LockID,AttributeType=S \ --key-
schema AttributeName=LockID,KeyType=HASH \ --
billing-mode PAY_PER_REQUEST
```

This command creates a DynamoDB table named **terraform-lock** with a primary key named **LockID** of type String and configures the table to use on-demand billing mode. Once the DynamoDB table is created, users can configure Terraform to use this table for state locking by specifying the **dynamodb_table** option in the backend configuration. For example, users can update their backend configuration to use DynamoDB-based locking as follows:

hclCopy code

terraform { backend "s3" { bucket = "my-terraform-state" key = "terraform.tfstate" region = "us-west-2" dynamodb_table = "terraform-lock" } }

In this configuration, the **dynamodb_table** option specifies the name of the DynamoDB table (**terraform-lock**) to use for state locking. Terraform automatically manages state locking using the specified DynamoDB table, ensuring that concurrent Terraform operations are serialized and coordinated to prevent conflicts and data corruption.

Another advanced state locking mechanism is Consul-based locking, which leverages HashiCorp Consul to implement distributed locking for Terraform state files. Consul is a distributed service mesh and key-value store designed for service discovery, configuration management, and distributed locking. By using Consul-based locking, users can achieve distributed coordination and consensus across multiple nodes or machines, enabling safe and reliable Terraform operations in large-scale or highly dynamic environments.

To configure Terraform to use Consul-based locking for state locking, users need to deploy a Consul cluster and configure Terraform to use this cluster for state locking. First, users can deploy a Consul cluster using the official Consul installation guide or using infrastructure as code tools like Terraform itself. Once the Consul cluster is deployed, users can configure Terraform to use this cluster for state locking by specifying the Consul address and session options in the backend configuration. For example, users can update their backend configuration to use Consul-based locking as follows:

hclCopy code

```
terraform { backend "consul" { address = "consul.example.com:8500" path = "terraform/state" scheme = "http" lock_timeout = "15s" unlock_timeout = "5s" } }
```

In this configuration, the **address** option specifies the address of the Consul cluster (**consul.example.com:8500**), the **path** option specifies the path within Consul's key-value store to store state data (**terraform/state**), and the **scheme** option specifies the communication protocol (**http**). Additionally, the **lock_timeout** and **unlock_timeout** options specify the timeout durations for acquiring and releasing locks, respectively. Terraform automatically manages state locking using the specified Consul cluster, ensuring that concurrent Terraform operations are serialized and coordinated to prevent conflicts and data corruption.

Overall, advanced state locking mechanisms play a crucial role in ensuring the integrity and consistency of Terraform state files in collaborative or distributed environments. By leveraging advanced state locking mechanisms like DynamoDB-based locking and Consul-based locking, users can prevent conflicts, data corruption, and race conditions when multiple users or processes interact with Terraform concurrently, improving the reliability and scalability of Terraform workflows.

Chapter 4: Infrastructure as Code Testing with Terraform

Unit testing Terraform code is a critical aspect of infrastructure as code (IaC) development, ensuring the reliability, correctness, and maintainability of Terraform configurations. Unit testing involves testing individual components or modules of Terraform code in isolation to validate their behavior and functionality. By writing unit tests for Terraform code, developers can identify and fix errors, prevent regressions, and promote code quality and consistency across deployments. Unit testing Terraform code typically involves writing test cases using testing frameworks like Terraform's built-in testing framework, Terratest, or external testing frameworks like Kitchen-Terraform.

To write unit tests for Terraform code using Terraform's built-in testing framework, developers can use the **terraform test** command, which runs Terraform configurations in a test environment and validates the results against expected outcomes. The **terraform test** command requires a test configuration file (usually named **test.tf**) that defines the test cases and their expected outcomes. For example, developers can define a test case that verifies the creation of an AWS EC2 instance with specific attributes using the following test configuration:

hclCopy code

```
provider "aws" { region = "us-west-2" } terraform {
required_providers { aws = { source = "hashicorp/aws"
version = "~> 3.0" } } } data "terraform_remote_state"
"network" { backend = "s3" config = { bucket = "my-
terraform-state" key = "network.tfstate" region = "us-
west-2" } } resource "aws_instance" "example" { ami =
"ami-0c55b159cbfafe1f0" instance_type = "t2.micro" }
output "instance_id" { value = aws_instance.example.id
} output "instance_ip" { value =
aws_instance.example.private_ip }
```

In this example, the test configuration defines a test
case that creates an AWS EC2 instance with the
specified AMI and instance type. The test configuration
also includes an output block that retrieves the instance
ID and private IP address for validation. To execute the
test case, developers can run the following command:
bashCopy code

```
terraform test
```

The **terraform test** command executes the test
configuration in a test environment, provisions the AWS
EC2 instance, and validates the instance attributes
against the expected outcomes defined in the test
configuration. If the test case passes, Terraform returns
a success message; otherwise, it displays error
messages indicating the failures.

Alternatively, developers can write unit tests for
Terraform code using Terratest, a testing framework
designed specifically for testing Terraform
configurations. Terratest provides a set of utilities and

helpers for writing and running automated tests for Terraform code, including infrastructure provisioning, validation, and cleanup. To write unit tests using Terratest, developers typically create Go test files that define test cases and use Terratest's testing APIs to interact with Terraform and validate the infrastructure.

For example, developers can write a Go test file named **terraform_test.go** that defines a test case for provisioning an AWS EC2 instance using Terraform and validates the instance attributes. The test case can use Terratest's **terraform.InitAndApply** function to initialize the Terraform configuration, provision the infrastructure, and retrieve the instance attributes for validation. Additionally, developers can use Terratest's assertion functions to verify the expected outcomes.

goCopy code

```
package test import ( "testing"
"github.com/gruntwork-
io/terratest/modules/terraform"
"github.com/stretchr/testify/assert" ) func
TestTerraformEC2Instance(t *testing.T) {
terraformOptions := &terraform.Options{ TerraformDir:
"../examples/ec2-instance", } defer
terraform.Destroy(t, terraformOptions)
terraform.InitAndApply(t, terraformOptions) instanceID
:= terraform.Output(t, terraformOptions,
"instance_id") instanceIP := terraform.Output(t,
terraformOptions, "instance_ip") assert.NotNil(t,
instanceID) assert.NotNil(t, instanceIP) }
```

In this example, the test case initializes the Terraform configuration located in the **../examples/ec2-instance** directory, applies the configuration to provision the AWS EC2 instance, and retrieves the instance ID and private IP address using Terratest's **terraform.Output** function. The test case then uses the **assert.NotNil** function from the **testify** library to assert that the instance ID and IP address are not nil, indicating that the infrastructure was provisioned successfully.

By writing unit tests for Terraform code using Terraform's built-in testing framework or external testing frameworks like Terratest, developers can ensure the reliability, correctness, and maintainability of their infrastructure deployments. Unit testing helps identify and fix errors early in the development process, promotes code quality and consistency, and increases confidence in the stability and reliability of Terraform configurations.

Integration testing infrastructure deployments is a crucial step in the development and deployment lifecycle of cloud-based applications, ensuring that the different components of the infrastructure work together as expected and meet the desired functionality and performance requirements. Integration testing involves testing the interactions and interfaces between various components of the infrastructure, such as servers, databases, networking components, and third-party services, to validate their interoperability and behavior in a production-like environment. By performing integration testing, organizations can identify and mitigate issues related to configuration errors, compatibility issues, communication failures, and performance bottlenecks before deploying the infrastructure to production environments.

To conduct integration testing of infrastructure deployments, organizations typically use automated testing frameworks and tools that facilitate the setup, execution, and validation of integration tests in a controlled and reproducible manner. One common approach to integration testing infrastructure deployments is to use a combination of infrastructure as code (IaC) tools like Terraform or AWS CloudFormation and testing frameworks like Terratest or AWS CDK to provision the infrastructure and validate its behavior. These tools enable developers and operations teams to define infrastructure configurations as code and automate the provisioning and testing processes, making it easier to set up and execute integration tests in different environments.

For example, organizations can use Terraform and Terratest to provision and test infrastructure deployments on AWS. To perform integration testing using Terraform and Terratest, developers first define the infrastructure configuration using Terraform's declarative configuration language, specifying the desired resources, configurations, and dependencies. Once the infrastructure configuration is defined, developers write test cases using Terratest to validate the behavior and functionality of the infrastructure components.

hclCopy code

```
provider "aws" { region = "us-west-2" } resource
"aws_instance" "example" { ami = "ami-
0c55b159cbfafe1f0" instance_type = "t2.micro" }
resource "aws_security_group" "example" { name =
"example" description = "Allow HTTP inbound traffic"
ingress { from_port = 80 to_port = 80 protocol = "tcp"
cidr_blocks = ["0.0.0.0/0"] } } output "instance_id" {
value = aws_instance.example.id }
```

In this example, the Terraform configuration defines an AWS EC2 instance and a security group that allows inbound HTTP traffic. To test the infrastructure deployment, developers write test cases using Terratest to validate the provisioning and configuration of the EC2 instance and security group.

goCopy code

```
package test import ( "testing"
"github.com/gruntwork-io/terratest/modules/aws"
"github.com/gruntwork-
io/terratest/modules/terraform"
"github.com/stretchr/testify/assert" ) func
TestTerraformIntegrationTest(t *testing.T) {
terraformOptions := &terraform.Options{ TerraformDir:
"../examples", } defer terraform.Destroy(t,
terraformOptions) terraform.InitAndApply(t,
terraformOptions) instanceID := terraform.Output(t,
terraformOptions, "instance_id") instance :=
aws.GetEc2Instance(t, instanceID, "us-west-2")
```

```
assert.Equal(t,    "t2.micro",    instance.InstanceType)
assert.Equal(t, "0.0.0.0/0", instance.PublicIpAddress) }
```

In this test case, the **TestTerraformIntegrationTest** function initializes the Terraform configuration located in the **../examples** directory, applies the configuration to provision the infrastructure, and retrieves the instance ID using Terratest's **terraform.Output** function. The test case then uses Terratest's **aws.GetEc2Instance** function to retrieve information about the provisioned EC2 instance and asserts that the instance type and public IP address match the expected values.

By using Terraform and Terratest for integration testing of infrastructure deployments, organizations can automate the testing process, validate the behavior and functionality of their infrastructure configurations, and ensure the reliability and correctness of their deployments in different environments. Integration testing enables organizations to identify and address issues early in the development lifecycle, reduce the risk of failures and downtime in production environments, and improve the overall quality and stability of their cloud-based applications.

Chapter 5: Terraform Workspaces and Environment Management

Managing multiple environments with Terraform workspaces is a fundamental aspect of infrastructure as code (IaC) development, enabling organizations to maintain separate, isolated environments for development, testing, staging, and production while using a single set of Terraform configurations. Terraform workspaces provide a mechanism for managing multiple state files and configurations within a single Terraform project, allowing users to switch between different environments seamlessly and manage their infrastructure deployments efficiently. By leveraging Terraform workspaces, organizations can streamline their development workflows, improve collaboration between teams, and maintain consistent infrastructure configurations across different environments.

To effectively manage multiple environments with Terraform workspaces, organizations typically adopt a naming convention and directory structure that reflects the different environments and their associated configurations. For example, organizations may organize their Terraform configurations into separate directories for each environment, such as **dev**, **qa**, **stage**, and **prod**, each containing environment-specific configuration files and state files. Additionally, organizations may use naming conventions to distinguish between different workspaces within each environment, such as **dev, test, uat**, and **prod**.

bashCopy code

```
# Create a new Terraform workspace for the dev environment terraform workspace new dev # Switch to the dev workspace terraform workspace select dev # Apply the Terraform configuration for the dev environment terraform apply # Create a new Terraform workspace for the prod environment terraform workspace new prod # Switch to the prod workspace terraform workspace select prod # Apply the Terraform configuration for the prod environment terraform apply
```

In this example, the **terraform workspace new** command is used to create a new workspace for each environment, such as **dev** and **prod**, while the **terraform workspace select** command is used to switch between different workspaces. Once the appropriate workspace is selected, the **terraform apply** command can be used to apply the Terraform configuration for the corresponding environment, provisioning and managing the infrastructure resources accordingly.

One of the key benefits of using Terraform workspaces is the ability to maintain separate state files for each environment, ensuring that changes made to one environment do not impact others. When a new workspace is created, Terraform automatically creates a separate state file for that workspace, allowing users to track and manage the state of their infrastructure deployments independently. This separation of state files helps minimize the risk of conflicts and synchronization issues between different environments, improving the stability and reliability of infrastructure deployments.

hclCopy code

terraform { backend "s3" { bucket = "my-terraform-state" key = "terraform.tfstate" region = "us-west-2" workspace_key_prefix = "env/" } }

In this example, the Terraform backend configuration specifies the **workspace_key_prefix** option, which prefixes the workspace name to the state file key in the S3 bucket. This ensures that each workspace has its own separate state file within the same bucket, facilitating isolation and management of state files for different environments.

Another advantage of using Terraform workspaces is the ability to parameterize configuration values and dynamically select values based on the current workspace. This allows users to define environment-specific configuration settings and variables within their Terraform configurations, making it easier to manage and deploy infrastructure across different environments.

hclCopy code

variable "environment" { type = string default = "dev" } resource "aws_instance" "example" { ami = var.environment == "prod" ? "ami-12345678" : "ami-abcdef12" instance_type = "t2.micro" }

In this example, the **environment** variable is used to specify the current environment, with a default value of

dev. The **aws_instance** resource uses a conditional expression to select the appropriate AMI based on the current environment, ensuring that the correct AMI is used for each environment.

Overall, managing multiple environments with Terraform workspaces offers significant benefits in terms of organization, isolation, and consistency of infrastructure deployments. By adopting best practices for managing workspaces, organizations can improve their development workflows, enhance collaboration between teams, and maintain reliable and consistent infrastructure configurations across different environments.

Environment isolation strategies are essential components of modern software development and infrastructure management, enabling organizations to maintain separate, distinct environments for different stages of the development lifecycle, such as development, testing, staging, and production. These strategies help ensure that changes made in one environment do not impact others, minimizing the risk of disruptions and providing a controlled and predictable environment for testing and deployment. Environment isolation is particularly crucial in cloud-based and distributed systems, where changes can have far-reaching consequences and where multiple teams may be working on different features simultaneously.

One common approach to environment isolation is to use separate infrastructure environments for each stage of the development lifecycle. For example, organizations may have dedicated environments for development, testing, staging, and production, each with its own set of infrastructure resources and configurations. This approach allows teams to work independently in their respective environments, without affecting others, and provides a clear separation of concerns between different stages of the development process.

To implement environment isolation using separate infrastructure environments, organizations can leverage infrastructure as code (IaC) tools like Terraform or AWS CloudFormation to define and provision the infrastructure resources for each environment. By codifying the infrastructure configurations, organizations can automate the provisioning process and ensure consistency across different environments, reducing the risk of configuration drift and human error.

bashCopy code

Create a new Terraform workspace for the dev environment terraform workspace new dev # Switch to the dev workspace terraform workspace select dev # Apply the Terraform configuration for the dev environment terraform apply # Create a new Terraform workspace for the prod environment terraform workspace new prod # Switch to the prod workspace terraform workspace select prod # Apply the Terraform configuration for the prod environment terraform apply

In this example, Terraform workspaces are used to manage separate environments for development (**dev**) and production (**prod**). Each environment has its own set of infrastructure configurations and state files, allowing teams to work independently and maintain isolation between environments. The **terraform apply** command is used to apply the Terraform configurations and provision the infrastructure resources for each environment.

Another approach to environment isolation is to use containerization technologies like Docker and Kubernetes to encapsulate and isolate application components within containers. Containers provide lightweight, portable environments that can be easily deployed and scaled across different stages of the development lifecycle. By containerizing applications and services, organizations can ensure consistency and repeatability in their deployments, regardless of the underlying infrastructure or environment.

bashCopy code

```
# Build the Docker image for the application docker build -t myapp:latest . # Run the Docker container locally for testing docker run -d -p 8080:80 myapp:latest # Push the Docker image to a container registry docker push myregistry/myapp:latest # Deploy the Docker container to a Kubernetes cluster kubectl apply -f deployment.yaml
```

In this example, Docker is used to build and run a containerized version of the application locally for testing purposes. The **docker build** command is used to build the Docker image, while the **docker run** command is used to run the container locally on port 8080. Once the container has been tested locally, the **docker push** command is used to push the Docker image to a container registry, such as Docker Hub or Amazon ECR. Finally, the **kubectl apply** command is used to deploy the Docker container to a Kubernetes cluster, where it can be scaled and managed in a production environment.

Additionally, organizations may implement environment isolation through network segmentation and access controls, restricting access to certain environments based on roles and permissions. By enforcing strict access controls and permissions, organizations can prevent unauthorized access to sensitive environments and data, reducing the risk of security breaches and compliance violations.

Overall, environment isolation strategies are essential for maintaining the integrity, reliability, and security of infrastructure deployments in modern software development environments. Whether through separate infrastructure environments, containerization, or access controls, organizations can ensure that changes are tested and deployed in controlled environments, minimizing the risk of disruptions and providing a stable and predictable environment for development and operations teams.

Chapter 6: Advanced Networking and Security with Terraform

VPC (Virtual Private Cloud) design patterns and best practices play a crucial role in architecting scalable, secure, and resilient cloud infrastructures. VPCs enable organizations to define and control a virtual network environment within their cloud provider's infrastructure, allowing them to isolate resources, control network traffic, and implement security policies. By adopting VPC design patterns and best practices, organizations can optimize their network architectures, enhance security posture, and improve the overall performance and reliability of their cloud-based applications and services.

One fundamental aspect of VPC design is the proper segmentation of network resources into subnets, which allows organizations to logically divide their network infrastructure and control the flow of traffic between different components. AWS provides a comprehensive set of CLI commands for managing VPC resources, including creating subnets, route tables, and network access control lists (NACLs).

bashCopy code

```
# Create a new VPC aws ec2 create-vpc --cidr-block
10.0.0.0/16 # Create public and private subnets aws
ec2 create-subnet --vpc-id <vpc-id> --cidr-block
10.0.1.0/24 aws ec2 create-subnet --vpc-id <vpc-id> --
```

cidr-block 10.0.2.0/24 # Create an internet gateway aws ec2 create-internet-gateway # Attach the internet gateway to the VPC aws ec2 attach-internet-gateway --vpc-id <vpc-id> --internet-gateway-id <internet-gateway-id> # Create a route table for public subnets aws ec2 create-route-table --vpc-id <vpc-id> # Create a route for internet traffic aws ec2 create-route --route-table-id <route-table-id> --destination-cidr-block 0.0.0.0/0 --gateway-id <internet-gateway-id> # Associate the route table with the public subnet aws ec2 associate-route-table --subnet-id <public-subnet-id> --route-table-id <route-table-id>

In this example, AWS CLI commands are used to create a new VPC, public and private subnets, an internet gateway, and route tables for routing internet traffic to the public subnets. By properly segmenting network resources into subnets and configuring routing tables, organizations can achieve network isolation and control the flow of traffic within their VPC.

Another essential aspect of VPC design is implementing security best practices to protect network resources from unauthorized access and malicious attacks. AWS provides a range of security features and services, such as security groups and network ACLs, for controlling inbound and outbound traffic to instances within a VPC.

```bash
bashCopy code
# Create a security group for web servers aws ec2 create-security-group --group-name web-sg --description "Security group for web servers" --vpc-id <vpc-id> # Authorize inbound traffic on port 80 (HTTP) aws ec2 authorize-security-group-ingress --group-id <web-sg-id> --protocol tcp --port 80 --cidr 0.0.0.0/0 # Create a security group for database servers aws ec2 create-security-group --group-name db-sg --description "Security group for database servers" --vpc-id <vpc-id> # Authorize inbound traffic on port 3306 (MySQL) from the web server security group aws ec2 authorize-security-group-ingress --group-id <db-sg-id> --protocol tcp --port 3306 --source-group <web-sg-id>
```

In this example, AWS CLI commands are used to create security groups for web servers and database servers and configure inbound traffic rules to allow HTTP traffic from any source to the web server security group and MySQL traffic from the web server security group to the database server security group. By implementing security groups and network ACLs, organizations can enforce network security policies and control access to resources within their VPC.

Additionally, organizations can leverage VPC peering and VPN connections to establish secure and private communication channels between VPCs and on-premises networks. VPC peering allows organizations to connect VPCs within the same AWS region, enabling them to share resources and communicate securely over the AWS network backbone.

bashCopy code

```
# Create a VPC peering connection aws ec2 create-vpc-peering-connection --vpc-id <vpc-id> --peer-vpc-id <peer-vpc-id> # Accept the VPC peering connection request aws ec2 accept-vpc-peering-connection --vpc-peering-connection-id <peering-connection-id>
```

In this example, AWS CLI commands are used to create a VPC peering connection between two VPCs and accept the peering connection request. Once the VPC peering connection is established, instances within the peered VPCs can communicate with each other securely and privately, without traversing the public internet.

Overall, adopting VPC design patterns and best practices is essential for building scalable, secure, and resilient cloud infrastructures. By properly segmenting network resources, implementing security controls, and establishing secure communication channels, organizations can optimize their network architectures and ensure the reliability and security of their cloud-based applications and services.

Implementing security policies with Terraform is a crucial aspect of managing infrastructure as code (IaC) and ensuring the security and compliance of cloud environments. Terraform provides a powerful and flexible platform for defining, deploying, and managing infrastructure resources, allowing organizations to codify security policies and best practices into their infrastructure configurations. By leveraging Terraform, organizations can automate the enforcement of security controls, maintain consistency across environments, and mitigate security risks effectively.

One common security policy that organizations often implement with Terraform is the enforcement of network security rules using security groups in cloud environments such as AWS or Azure. Security groups act as virtual firewalls that control inbound and outbound traffic to instances within a specific VPC or subnet. Using Terraform, organizations can define security group rules to restrict access to instances based on IP addresses, ports, and protocols.

hclCopy code

```
resource "aws_security_group" "web" { name = "web-sg" description = "Security group for web servers" vpc_id = aws_vpc.default.id ingress { from_port = 80 to_port = 80 protocol = "tcp" cidr_blocks = ["0.0.0.0/0"] } egress { from_port = 0 to_port = 65535 protocol = "tcp" cidr_blocks = ["0.0.0.0/0"] } }
```

In this example, Terraform is used to define an AWS security group for web servers. The security group allows inbound traffic on port 80 (HTTP) from any source (**0.0.0.0/0**) and allows all outbound traffic. By defining security group rules in Terraform configurations, organizations can ensure that instances are only accessible from authorized sources and that traffic flows are restricted according to their security policies.

Another common security policy that organizations implement with Terraform is the encryption of data at rest using AWS Key Management Service (KMS) or other encryption mechanisms. Terraform allows organizations to define encryption settings for various resources, such as Amazon EBS volumes, Amazon S3 buckets, and AWS RDS databases, ensuring that sensitive data is protected from unauthorized access.

hclCopy code

```
resource "aws_ebs_volume" "example" { availability_zone = "us-east-1a" size = 100 encrypted = true kms_key_id = "arn:aws:kms:us-east-1:123456789012:key/abcd1234-abcd-1234-abcd-1234567890ab" }
```

In this example, Terraform is used to define an encrypted Amazon EBS volume using a specific AWS KMS key. By specifying the **encrypted** attribute and the **kms_key_id**, Terraform ensures that the EBS volume is encrypted with the specified KMS key, providing an additional layer of security for data stored on the volume.

Furthermore, organizations can use Terraform to implement identity and access management (IAM) policies to control access to AWS resources and enforce least privilege principles. IAM policies define permissions for users, groups, and roles, specifying which actions they are allowed or denied to perform on AWS resources.

hclCopy code

```
resource "aws_iam_policy" "example" { name = "example-policy" description = "Example IAM policy" policy = <<EOF { "Version": "2012-10-17", "Statement": [ { "Effect": "Allow", "Action": [ "s3:GetObject", "s3:PutObject" ], "Resource": "arn:aws:s3:::example-bucket/*" } ] } EOF }
```

In this example, Terraform is used to define an IAM policy that grants permissions to read and write objects in an Amazon S3 bucket. By specifying the **Effect**, **Action**, and **Resource** attributes, organizations can define granular permissions for AWS resources, ensuring that only authorized users, groups, or roles have access to sensitive data or operations.

Overall, implementing security policies with Terraform is essential for ensuring the security and compliance of cloud environments. By codifying security controls into infrastructure configurations, organizations can automate the enforcement of security policies, maintain consistency across environments, and reduce the risk of security breaches and compliance violations.

Chapter 7: CI/CD Pipelines for Terraform Automation

Building CI/CD pipelines for Terraform is an essential practice for organizations aiming to automate the deployment and management of infrastructure as code (IaC) changes. CI/CD, which stands for Continuous Integration/Continuous Deployment, enables teams to streamline the process of testing, building, and deploying infrastructure changes in a reliable and efficient manner. By incorporating Terraform into CI/CD pipelines, organizations can ensure consistency, repeatability, and scalability in their infrastructure deployments.

One of the first steps in building CI/CD pipelines for Terraform is setting up version control for infrastructure code using a version control system like Git. Version control allows teams to track changes to their Terraform configurations over time, collaborate effectively, and revert to previous versions if needed. With Git, teams can use commands like **git clone**, **git add**, **git commit**, and **git push** to manage their Terraform code repositories.

bashCopy code

Clone the existing Terraform repository git clone <repository-url> # Add Terraform configuration files to the repository git add . # Commit the changes git commit -m "Add initial Terraform configurations"

Push the changes to the remote repository git push origin master

In this example, the **git clone** command is used to clone an existing Terraform repository from a remote Git server. The **git add** command adds Terraform configuration files to the repository, while the **git commit** command commits the changes with a descriptive message. Finally, the **git push** command pushes the changes to the remote repository, making them accessible to other team members and CI/CD pipelines.

Once version control is set up, organizations can integrate Terraform into their CI/CD pipelines using popular CI/CD platforms such as Jenkins, GitLab CI/CD, or AWS CodePipeline. These platforms provide tools and services for automating the process of testing, building, and deploying Terraform configurations, ensuring that changes are validated and deployed consistently across environments.

For example, organizations can use Jenkins, an open-source automation server, to create CI/CD pipelines for Terraform. Jenkins pipelines are defined using Jenkinsfile, a Groovy-based script that specifies the steps to be executed during the CI/CD process. Below is an example of a Jenkinsfile that defines a CI/CD pipeline for Terraform:

groovyCopy code

pipeline { agent any stages { stage('Checkout') { steps { git 'https://github.com/organization/terraform.git' }

```
} stage('Terraform Plan') { steps { sh 'terraform init' sh
'terraform plan -out=tfplan' } } stage('Terraform
Apply') { steps { sh 'terraform apply tfplan' } } } }
```

In this Jenkinsfile, the pipeline consists of three stages: Checkout, Terraform Plan, and Terraform Apply. In the Checkout stage, the Terraform code repository is cloned from the remote Git server. In the Terraform Plan stage, Terraform is initialized, and a plan is generated using the **terraform plan** command. Finally, in the Terraform Apply stage, the plan is applied using the **terraform apply** command, deploying the changes to the target environment.

Similarly, organizations can use AWS CodePipeline, a fully managed CI/CD service, to build automated pipelines for Terraform on AWS. AWS CodePipeline allows organizations to define custom pipelines with multiple stages, including source, build, test, and deploy stages. Below is an example of a CodePipeline configuration file that defines a CI/CD pipeline for Terraform:

```
yamlCopy code
version: 0.2 phases: install: commands: - yum
install -y unzip - curl -o terraform.zip
https://releases.hashicorp.com/terraform/0.14.10/te
rraform_0.14.10_linux_amd64.zip        -        unzip
terraform.zip - mv terraform /usr/local/bin/
pre_build: commands: - terraform init build:
commands:    -    terraform    plan    -out=tfplan
post_build: commands: - terraform apply tfplan
```

In this CodePipeline configuration file, Terraform is installed and initialized in the install and pre_build phases, respectively. The Terraform plan is generated in the build phase using the **terraform plan** command, and the changes are applied in the post_build phase using the **terraform apply** command. By defining these phases and commands in the CodePipeline configuration file, organizations can automate the process of testing, building, and deploying Terraform configurations on AWS.

Overall, building CI/CD pipelines for Terraform is crucial for automating the deployment and management of infrastructure as code changes. By integrating Terraform into CI/CD pipelines, organizations can ensure consistency, reliability, and scalability in their infrastructure deployments, accelerating the delivery of applications and services while minimizing the risk of errors and downtime.

Automating deployment pipelines with Terraform is a fundamental practice for modern software development and infrastructure management. As organizations embrace DevOps methodologies and seek to accelerate the delivery of software applications, the automation of deployment pipelines becomes essential for streamlining the process of building, testing, and deploying changes to production environments. Terraform, with its infrastructure as code (IaC) capabilities, provides a robust framework for automating deployment pipelines, enabling organizations to define, version, and manage their infrastructure configurations as code.

One of the key benefits of automating deployment pipelines with Terraform is the ability to codify infrastructure changes and deploy them consistently across different environments. By defining infrastructure configurations as Terraform code, organizations can capture their infrastructure requirements in a declarative manner, specifying the desired state of the infrastructure rather than imperatively defining the steps to achieve that state. This declarative approach to infrastructure management ensures that deployments are predictable, repeatable, and version-controlled, reducing the risk of configuration drift and ensuring consistency across environments.

To automate deployment pipelines with Terraform, organizations typically use continuous integration and continuous deployment (CI/CD) tools such as Jenkins, GitLab CI/CD, or AWS CodePipeline. These tools allow organizations to define pipelines that automate the process of building, testing, and deploying infrastructure changes, integrating Terraform into the CI/CD workflow seamlessly.

For example, organizations can use Jenkins, an open-source automation server, to create CI/CD pipelines for Terraform. Jenkins pipelines are defined using Jenkinsfile, a Groovy-based script that specifies the stages and steps of the CI/CD process. Below is an example of a Jenkinsfile that defines a CI/CD pipeline for Terraform:

groovyCopy code

```
pipeline { agent any stages { stage('Checkout') { steps
{ git 'https://github.com/organization/terraform.git' }
} stage('Terraform Plan') { steps { sh 'terraform init' sh
'terraform plan -out=tfplan' } } stage('Terraform
Apply') { steps { sh 'terraform apply tfplan' } } } }
```

In this Jenkinsfile, the pipeline consists of three stages: Checkout, Terraform Plan, and Terraform Apply. In the Checkout stage, the Terraform code repository is cloned from the remote Git server. In the Terraform Plan stage, Terraform is initialized, and a plan is generated using the **terraform plan** command. Finally, in the Terraform Apply stage, the plan is applied using the **terraform apply** command, deploying the changes to the target environment.

Similarly, organizations can use AWS CodePipeline, a fully managed CI/CD service, to automate deployment pipelines for Terraform on AWS. AWS CodePipeline allows organizations to define custom pipelines with multiple stages, including source, build, test, and deploy stages. Below is an example of a CodePipeline configuration file that defines a CI/CD pipeline for Terraform:

yamlCopy code

```
version: 0.2 phases: install: commands: - yum
install -y unzip - curl -o terraform.zip
https://releases.hashicorp.com/terraform/0.14.10/te
```

rraform_0.14.10_linux_amd64.zip - unzip terraform.zip - mv terraform /usr/local/bin/ pre_build: commands: - terraform init build: commands: - terraform plan -out=tfplan post_build: commands: - terraform apply tfplan

In this CodePipeline configuration file, Terraform is installed and initialized in the install and pre_build phases, respectively. The Terraform plan is generated in the build phase using the **terraform plan** command, and the changes are applied in the post_build phase using the **terraform apply** command. By defining these phases and commands in the CodePipeline configuration file, organizations can automate the process of testing, building, and deploying Terraform configurations on AWS.

Overall, automating deployment pipelines with Terraform enables organizations to streamline the process of deploying infrastructure changes, reduce manual intervention, and increase the speed and reliability of software delivery. By integrating Terraform into CI/CD pipelines, organizations can achieve greater efficiency, agility, and consistency in their infrastructure deployments, ultimately accelerating the delivery of value to customers and stakeholders.

Chapter 8: Scaling Infrastructure with Terraform and AWS Auto Scaling

Dynamic scaling strategies are pivotal components of modern cloud infrastructure management, facilitating the automatic adjustment of resources to meet varying workload demands. In dynamic scaling, resources such as compute instances, databases, and storage are provisioned or deprovisioned dynamically based on real-time metrics, ensuring optimal performance, cost efficiency, and reliability. Cloud providers offer various tools and services to implement dynamic scaling strategies, empowering organizations to scale their infrastructure elastically in response to fluctuating workloads. One of the widely adopted approaches to dynamic scaling is auto-scaling, which enables organizations to automatically adjust the number of compute instances or resources in response to changes in demand. In AWS, auto-scaling can be configured using the Auto Scaling service, allowing organizations to define scaling policies based on metrics such as CPU utilization, network traffic, or custom application metrics. For example, organizations can use the AWS Management Console or AWS CLI to create an auto-scaling group and define scaling policies:
bashCopy code

```
aws autoscaling create-auto-scaling-group --auto-
scaling-group-name my-asg \ --launch-template
LaunchTemplateId=my-launch-template-
id,Version=my-launch-template-version \ --min-size 2
--max-size 10 --desired-capacity 2
```

In this command, an auto-scaling group named "my-asg" is created with a minimum size of 2 instances, a maximum size of 10 instances, and a desired capacity of 2 instances. Additionally, organizations can define scaling policies to scale the auto-scaling group based on specific conditions:

bashCopy code

```
aws autoscaling put-scaling-policy --policy-name my-
scaling-policy \ --auto-scaling-group-name my-asg --
scaling-adjustment 2 --adjustment-type
ChangeInCapacity
```

This command creates a scaling policy named "my-scaling-policy" that adds 2 instances to the auto-scaling group when triggered.

Another dynamic scaling strategy involves leveraging serverless computing platforms such as AWS Lambda, which automatically scales compute resources in response to incoming requests or events. With AWS Lambda, organizations can execute code without provisioning or managing servers, allowing them to focus on building and deploying applications without worrying about infrastructure management. AWS Lambda automatically scales the underlying infrastructure based on the number of incoming requests, ensuring that resources are allocated efficiently to handle workload spikes.

Furthermore, container orchestration platforms like Amazon Elastic Kubernetes Service (Amazon EKS) provide dynamic scaling capabilities for containerized workloads. Organizations can define horizontal pod autoscalers (HPAs) to automatically adjust the number of pods in a Kubernetes cluster based on metrics such as CPU utilization or memory usage. By leveraging HPAs, organizations can ensure that their containerized applications can scale seamlessly to meet changing demand, optimizing resource utilization and enhancing application performance.

In addition to auto-scaling, organizations can implement dynamic scaling strategies using predictive scaling, which uses machine learning algorithms to forecast future workload demand and adjust resources preemptively. AWS offers services like Amazon EC2 Auto Scaling Predictive Scaling, which uses historical data and machine learning to forecast future demand and provision resources accordingly. By proactively scaling resources based on predicted demand, organizations can improve responsiveness and cost efficiency while minimizing the risk of under-provisioning or over-provisioning.

Moreover, organizations can implement dynamic scaling strategies at the database level by leveraging services like Amazon RDS (Relational Database Service) or Amazon Aurora. These services offer features such as Read Replicas and Multi-AZ deployments, allowing organizations to dynamically scale database capacity to accommodate changes in workload demand. For example, organizations can use Amazon RDS Multi-AZ deployments to automatically provision and maintain synchronous standby replicas in different Availability Zones, ensuring high availability and fault tolerance.

Furthermore, organizations can implement dynamic scaling strategies for storage resources using services like Amazon S3 (Simple Storage Service) or Amazon EBS (Elastic Block Store). Amazon S3 automatically scales to accommodate any amount of data, allowing organizations to store and retrieve data reliably and cost-effectively. Similarly, Amazon EBS provides scalable block storage volumes that can be dynamically resized to meet changing capacity requirements, ensuring that applications have access to the storage resources they need.

In summary, dynamic scaling strategies are essential for optimizing cloud infrastructure performance, cost efficiency, and reliability. By leveraging tools and services offered by cloud providers like AWS, organizations can implement auto-scaling, serverless computing, container orchestration, predictive scaling, and dynamic scaling for databases and storage resources. These strategies enable organizations to scale their infrastructure dynamically in response to fluctuating workloads, ensuring optimal performance and cost efficiency while minimizing manual intervention.

Implementing auto scaling with Terraform is a critical aspect of modern cloud infrastructure management, enabling organizations to dynamically adjust their resources to match fluctuating demand patterns automatically. Terraform, a popular infrastructure as code (IaC) tool, allows users to define and provision infrastructure resources using declarative configuration files, making it well-suited for automating the setup of auto-scaling environments. Auto scaling ensures that organizations can efficiently allocate resources based on real-time metrics such as CPU utilization, memory usage, or incoming network traffic, optimizing performance while minimizing costs.

One of the key components of implementing auto scaling with Terraform is defining auto scaling groups (ASGs), which are logical groupings of EC2 instances or other resources that can automatically scale based on predefined conditions. Terraform provides a dedicated resource type for creating auto scaling groups, allowing users to specify various parameters such as minimum and maximum instance counts, desired capacity, launch configurations, and scaling policies.

```
hclCopy code
resource "aws_autoscaling_group" "example" { name
= "example-asg" max_size = 5 min_size = 1
desired_capacity = 2 launch_configuration =
```

aws_launch_configuration.example.name

vpc_zone_identifier = [aws_subnet.example.id] }

In this Terraform configuration snippet, an auto scaling group named "example-asg" is defined with a minimum size of 1 instance, a maximum size of 5 instances, and a desired capacity of 2 instances. The **launch_configuration** parameter references an existing launch configuration resource, which specifies the configuration details for the EC2 instances launched by the auto scaling group. The **vpc_zone_identifier** parameter specifies the subnet IDs where the instances will be launched.

Additionally, Terraform allows users to define scaling policies for auto scaling groups, which specify the conditions under which the group should scale in or out. Scaling policies can be based on various metrics such as CPU utilization, network traffic, or custom CloudWatch metrics. Below is an example of how to define a scaling policy for an auto scaling group:

hclCopy code

```
resource "aws_autoscaling_policy" "example" { name = "example-policy" scaling_adjustment = 1 adjustment_type = "ChangeInCapacity" cooldown = 300 autoscaling_group_name = aws_autoscaling_group.example.name
step_adjustment { metric_interval_lower_bound = 0 scaling_adjustment = 1 } }
```

In this Terraform configuration, a scaling policy named "example-policy" is defined with a scaling adjustment of 1 instance and a cooldown period of 300 seconds. The **autoscaling_group_name** parameter specifies the auto scaling group to which the policy applies. Additionally, a step adjustment is defined to increment the desired capacity by 1 instance when certain conditions are met.

Furthermore, Terraform enables users to monitor and manage auto scaling resources using the AWS provider and various data sources. For example, the **aws_autoscaling_group** data source can be used to retrieve information about existing auto scaling groups, allowing users to dynamically reference and interact with them in their Terraform configurations.

hclCopy code

```
data "aws_autoscaling_group" "example" { name = "example-asg" }
```

This data source retrieves information about the auto scaling group named "example-asg" and makes it available for use within Terraform configurations. Users can then access attributes of the auto scaling group, such as its ARN or desired capacity, and use them in other parts of their Terraform code.

In addition to EC2 instances, Terraform supports auto scaling for other AWS resources such as ECS services, DynamoDB tables, and Aurora database clusters. This allows organizations to implement auto scaling across a wide range of AWS services, ensuring that their infrastructure can adapt to changing workload demands effectively.

Overall, implementing auto scaling with Terraform empowers organizations to build resilient, cost-effective, and scalable cloud infrastructure environments. By leveraging Terraform's declarative syntax and powerful automation capabilities, users can define, deploy, and manage auto scaling resources with ease, ensuring that their applications can scale seamlessly to meet dynamic workload requirements.

Chapter 9: Advanced Data Management in AWS with Terraform

Database deployment strategies with Terraform are crucial for efficiently provisioning and managing database resources in cloud environments. Terraform, as a versatile infrastructure as code (IaC) tool, provides a robust framework for automating the deployment of various database solutions, including relational databases like Amazon RDS, NoSQL databases like Amazon DynamoDB, and managed database services such as Amazon Aurora. These deployment strategies encompass a range of techniques and best practices to ensure that databases are provisioned, configured, and maintained effectively to meet the requirements of modern applications.

One of the fundamental aspects of database deployment with Terraform is defining database resources using Terraform configuration files. These files specify the desired state of the database infrastructure, including attributes such as database engine, instance type, storage size, backup settings, and access controls. For example, to provision an Amazon RDS MySQL database using Terraform, users can create a Terraform configuration file like the following:

hclCopy code

resource "aws_db_instance" "example" { identifier = "example-db" engine = "mysql" instance_class = "db.t2.micro" allocated_storage = 10 storage_type =

"gp2" username = "admin" password = "examplePassword123" publicly_accessible = true }

In this Terraform configuration snippet, an Amazon RDS MySQL database instance named "example-db" is defined with the specified engine, instance type, storage size, and access credentials. The **publicly_accessible** parameter allows the database to be accessed over the internet, if necessary. Once the configuration file is defined, users can apply the configuration using the **terraform apply** command to provision the database resources in their AWS account:

bashCopy code

terraform apply

This command instructs Terraform to analyze the configuration files, plan the changes to be made, and apply those changes to the infrastructure. Terraform will communicate with the AWS API to create the specified database resources, and once the process is complete, users will have a fully provisioned database instance ready for use.

In addition to provisioning databases, Terraform allows users to manage database configurations and settings using resource attributes and variables. This enables organizations to customize database configurations according to their specific requirements, such as enabling multi-AZ deployment, configuring backup retention policies, and setting up encryption options. For example, to enable automated backups for an Amazon RDS instance, users can specify the **backup_retention_period** attribute in the Terraform configuration:

hclCopy code

```
resource "aws_db_instance" "example" { # Other
attributes... backup_retention_period = 7 }
```

This configuration ensures that automated backups are retained for a period of 7 days, providing data durability and recoverability in the event of system failures or data corruption.

Furthermore, Terraform supports the management of database access controls and security settings, allowing users to define user accounts, database permissions, and network access rules. For example, users can use Terraform to create database users and grant them specific privileges on the database objects:

hclCopy code

```
resource "aws_db_instance" "example" { # Other
attributes... tags = { Name = "example-db" } } resource
"aws_db_instance_role_association" "example" {
db_instance_identifier = aws_db_instance.example.id
feature_name = "s3Import" role_arn =
aws_iam_role.example.arn }
```

In this configuration, a role association is established between the RDS instance and an IAM role, allowing the instance to access Amazon S3 for data import operations. This demonstrates how Terraform can be used to manage database permissions and integrate database resources with other AWS services.

Moreover, Terraform supports the management of database lifecycle operations such as scaling, monitoring, and maintenance. For example, users can define scaling policies to automatically adjust the capacity of database instances based on performance metrics or workload demands. Similarly, Terraform can be used to configure database monitoring and logging settings, enabling organizations to monitor database performance and diagnose issues effectively.

Additionally, Terraform integrates seamlessly with infrastructure orchestration tools and continuous integration/continuous deployment (CI/CD) pipelines, allowing users to automate the deployment and management of database resources as part of their application deployment workflows. By incorporating database deployment into automated CI/CD pipelines, organizations can streamline the software delivery process and ensure consistency and reliability across their infrastructure environments.

In summary, database deployment strategies with Terraform play a pivotal role in modern cloud infrastructure management, enabling organizations to provision, configure, and manage database resources effectively. By leveraging Terraform's infrastructure as code capabilities, users can define database configurations, manage access controls, automate lifecycle operations, and integrate databases into their application deployment workflows. This facilitates the adoption of agile, scalable, and cost-effective database solutions that meet the evolving needs of modern applications and business requirements.

Data encryption and compliance management are critical aspects of modern IT infrastructure management, especially in the context of cloud computing environments where sensitive data is often stored and processed. With the increasing emphasis on data privacy and regulatory requirements such as GDPR, HIPAA, and PCI DSS, organizations need robust solutions to protect their data and ensure compliance with relevant regulations. Next, we will explore how to leverage encryption techniques and compliance management practices using Terraform, a popular infrastructure as code (IaC) tool, to secure data and meet regulatory requirements effectively.

One of the key components of data encryption and compliance management is encrypting data at rest and in transit to protect it from unauthorized access. Terraform provides native support for configuring encryption settings for various AWS services, allowing users to enable encryption for data stored in databases, storage volumes, and communication channels. For example, to enable encryption for an Amazon S3 bucket using Terraform, users can define the encryption settings in their Terraform configuration files:

hclCopy code

```
resource "aws_s3_bucket" "example" { bucket = "example-bucket" acl = "private" server_side_encryption_configuration { rule { apply_server_side_encryption_by_default { sse_algorithm = "AES256" } } } }
```

In this Terraform configuration snippet, an Amazon S3 bucket named "example-bucket" is defined with encryption settings configured to use the AES256 encryption algorithm by default. This ensures that all objects stored in the S3 bucket are encrypted at rest, providing an additional layer of security to protect sensitive data.

Furthermore, Terraform allows users to manage encryption keys and access controls for encrypted resources, ensuring that only authorized users and services can access encrypted data. For example, users can create and manage AWS Key Management Service (KMS) keys using Terraform to encrypt and decrypt data stored in AWS services such as S3, EBS, and RDS. The following Terraform configuration demonstrates how to create a KMS key with the necessary permissions:

hclCopy code

```
resource "aws_kms_key" "example" { description = "Example KMS key" deletion_window_in_days = 10 policy = <<-POLICY { "Version": "2012-10-17", "Id": "key-default-1", "Statement": [ { "Sid": "Enable IAM User Permissions", "Effect": "Allow", "Principal": { "AWS": "arn:aws:iam::123456789012:root" }, "Action": "kms:*", "Resource": "*" } ] } POLICY }
```

In this configuration, a KMS key named "example" is defined with a description and a deletion window of 10 days. The key policy grants permissions to the root IAM user to perform all KMS operations on the key, ensuring that the necessary permissions are in place for encrypting and decrypting data using the key.

Moreover, Terraform facilitates the integration of encryption and compliance management practices into continuous integration/continuous deployment (CI/CD) pipelines, enabling organizations to automate the deployment and management of encrypted resources as part of their application deployment workflows. By incorporating encryption and compliance management into automated CI/CD pipelines, organizations can ensure that security and compliance requirements are consistently enforced throughout the software development lifecycle.

Additionally, Terraform supports the implementation of compliance management controls and auditing mechanisms to monitor and enforce compliance with regulatory requirements. For example, users can define AWS Config rules using Terraform to assess the compliance of AWS resources against predefined configuration rules and remediate non-compliant resources automatically. The following Terraform configuration demonstrates how to define an AWS Config rule to enforce encryption settings for S3 buckets:

hclCopy code

```
resource "aws_config_config_rule" "example" { name = "s3-bucket-encryption-compliance" source { owner = "AWS" source_identifier = "S3_BUCKET_ENCRYPTION_ENABLED" } }
```

In this configuration, an AWS Config rule named "s3-bucket-encryption-compliance" is defined to check whether encryption is enabled for S3 buckets. If a non-compliant S3 bucket is detected, AWS Config can trigger a remediation action to enable encryption automatically, ensuring that all S3 buckets comply with the encryption requirements.

In summary, data encryption and compliance management are essential components of modern IT infrastructure management, particularly in cloud environments where data security and regulatory compliance are paramount. By leveraging Terraform's capabilities, organizations can implement robust encryption techniques, manage encryption keys and access controls, integrate encryption and compliance management into CI/CD pipelines, and enforce compliance with regulatory requirements effectively. This enables organizations to protect their sensitive data, mitigate security risks, and demonstrate compliance with relevant regulations, thereby building trust with customers and stakeholders.

Chapter 10: Monitoring and Logging Infrastructure with Terraform

Setting up monitoring solutions with Terraform is crucial for ensuring the health, performance, and security of cloud infrastructure and applications. Terraform, as an infrastructure as code (IaC) tool, allows organizations to define and deploy monitoring resources and configurations programmatically, enabling consistent and scalable monitoring across distributed environments. Next, we will explore how to leverage Terraform to set up monitoring solutions using popular monitoring tools such as Prometheus, Grafana, and AWS CloudWatch, as well as integrations with third-party monitoring services.

One of the key aspects of setting up monitoring solutions with Terraform is defining monitoring resources and configurations in Terraform configuration files. These files specify the desired state of monitoring infrastructure, including metrics, alerts, dashboards, and integrations with monitoring tools. For example, to set up monitoring with Prometheus and Grafana using Terraform, users can define the necessary resources in their Terraform configuration files:

hclCopy code

```
# Define Prometheus server resource "aws_instance" "prometheus_server" { # Instance configuration... } # Define Grafana server resource "aws_instance"
```

"grafana_server" { # Instance configuration... } # Define Prometheus and Grafana security group rules resource "aws_security_group_rule" "prometheus_ingress" { # Security group rule configuration... } resource "aws_security_group_rule" "grafana_ingress" { # Security group rule configuration... }

In this Terraform configuration snippet, AWS EC2 instances for Prometheus and Grafana servers are defined along with security group rules to allow inbound traffic to the respective servers. Once the configuration files are defined, users can apply the configuration using the **terraform apply** command to provision the monitoring infrastructure in their AWS account:

bashCopy code

terraform apply

This command instructs Terraform to analyze the configuration files, plan the changes to be made, and apply those changes to the infrastructure. Terraform will communicate with the AWS API to create the specified EC2 instances and security group rules, and once the process is complete, users will have Prometheus and Grafana servers deployed and ready for use.

Furthermore, Terraform allows users to configure monitoring resources and integrations with monitoring tools to collect, store, and visualize metrics from various sources such as servers, containers, databases, and applications. For example, users can use Terraform to define Prometheus configuration files to scrape metrics from target endpoints and Grafana dashboards to visualize the collected metrics. The following Terraform configuration demonstrates how to define Prometheus and Grafana configurations:

hclCopy code

```
# Define Prometheus configuration resource "aws_s3_bucket_object" "prometheus_config" { # Prometheus configuration file... } # Define Grafana dashboard resource "aws_s3_bucket_object" "grafana_dashboard" { # Grafana dashboard file... }
```

In this configuration, Prometheus configuration file and Grafana dashboard file are defined as AWS S3 bucket objects. These files contain the necessary configurations and definitions to enable Prometheus to scrape metrics from target endpoints and Grafana to visualize the collected metrics in interactive dashboards.

Moreover, Terraform supports integrations with AWS CloudWatch, a fully managed monitoring and observability service provided by AWS, to collect and visualize metrics, logs, and events from AWS resources and applications. Users can use Terraform to define CloudWatch alarms, dashboards, and metric filters to monitor the performance and health of AWS services and applications. For example, users can define CloudWatch alarm configuration using Terraform to monitor CPU utilization of EC2 instances:

hclCopy code

```
# Define CloudWatch alarm resource "aws_cloudwatch_metric_alarm" "cpu_utilization_alarm" { # Alarm configuration... }
```

In this configuration, a CloudWatch alarm named "cpu_utilization_alarm" is defined to monitor the CPU utilization metric of EC2 instances. If the CPU utilization exceeds the specified threshold, the alarm will trigger an action, such as sending a notification or executing an automated remediation action.

Additionally, Terraform enables the integration of third-party monitoring services and tools with cloud infrastructure to extend monitoring capabilities and provide insights into application performance and availability. For example, users can use Terraform to configure integrations with Datadog, New Relic, or Splunk to collect and analyze metrics, logs, and traces from cloud-native applications and microservices. This allows organizations to leverage the capabilities of third-party monitoring solutions to gain deeper insights into application performance, identify bottlenecks and issues, and optimize resource utilization.

In summary, setting up monitoring solutions with Terraform is essential for organizations to ensure the reliability, performance, and security of their cloud infrastructure and applications. By leveraging Terraform's infrastructure as code capabilities, users can define and deploy monitoring resources and configurations programmatically, enabling consistent and scalable monitoring across distributed environments. Whether it's configuring monitoring with Prometheus, Grafana, and AWS CloudWatch or integrating with third-party monitoring services, Terraform provides a flexible and powerful platform for implementing comprehensive monitoring solutions that meet the needs of modern cloud-native applications and infrastructure.

Centralized logging and analysis strategies are vital components of modern IT infrastructure management, enabling organizations to aggregate, store, and analyze logs from various sources to gain insights into system behavior, diagnose issues, and ensure compliance with regulatory requirements. With the increasing complexity and scale of cloud-based environments, centralized logging becomes indispensable for effective monitoring, troubleshooting, and security incident response. Next, we will explore how to implement centralized logging and analysis strategies using Terraform, a popular infrastructure as code (IaC) tool, along with other complementary technologies such as Amazon CloudWatch Logs, Elasticsearch, and Fluentd.

The first step in setting up centralized logging is to define the logging infrastructure and configurations in Terraform configuration files. Terraform allows users to provision and configure logging resources such as log groups, log streams, and retention policies programmatically, ensuring consistency and repeatability across environments. For example, to create a log group in Amazon CloudWatch Logs using Terraform, users can define the following Terraform configuration:

hclCopy code

```
# Define CloudWatch Logs log group
resource "aws_cloudwatch_log_group" "example_log_group" {
name = "/example" }
```

In this Terraform configuration snippet, an Amazon CloudWatch Logs log group named "/example" is defined, providing a centralized location to store logs from various sources. Once the Terraform configuration files are defined, users can apply the configuration using the **terraform apply** command to provision the logging infrastructure:

bashCopy code

terraform apply

This command instructs Terraform to analyze the configuration files, plan the changes to be made, and apply those changes to the infrastructure. Terraform communicates with the AWS API to create the specified CloudWatch Logs log group, ensuring that the logging infrastructure is provisioned according to the defined configuration.

Furthermore, Terraform enables users to configure log streams and retention policies to control the flow of logs and manage log retention duration. For example, users can define log stream configurations to route logs from different sources to specific log streams within the log group:

hclCopy code

Define CloudWatch Logs log stream resource "aws_cloudwatch_log_stream" "example_log_stream" { name = "example-stream" log_group_name = aws_cloudwatch_log_group.example_log_group.name }

In this configuration, a CloudWatch Logs log stream named "example-stream" is defined within the specified log group, allowing users to organize logs from different sources into separate streams for easier management and analysis. Additionally, users can define retention policies to specify the duration for which logs should be retained in the log group:

hclCopy code

```
# Define CloudWatch Logs retention policy resource "aws_cloudwatch_log_retention_policy" "example_retention_policy" { log_group_name = aws_cloudwatch_log_group.example_log_group.name retention_in_days = 30 }
```

In this configuration, a retention policy is defined to retain logs in the log group for 30 days, ensuring that logs are retained for the specified duration before being automatically deleted.

Moreover, Terraform supports integrations with other logging and analysis tools such as Elasticsearch, Fluentd, and Kibana to provide more advanced logging and analysis capabilities. For example, users can use Terraform to deploy and configure an Elasticsearch cluster and Fluentd agents to ingest logs from various sources and store them in Elasticsearch for indexing and analysis:

hclCopy code

```
# Define Elasticsearch domain resource "aws_elasticsearch_domain" "example_domain" { domain_name = "example-domain" elasticsearch_version = "7.10" cluster_config { instance_type = "t2.small.elasticsearch" } } # Define
```

Fluentd configuration resource "aws_instance" "fluentd_agent" { # Instance configuration... }

In this configuration, an Elasticsearch domain named "example-domain" is defined, along with a Fluentd agent deployed on an AWS EC2 instance to collect and forward logs to Elasticsearch for indexing and analysis.

In summary, centralized logging and analysis strategies are essential for effective monitoring, troubleshooting, and security incident response in modern IT environments. By leveraging Terraform's infrastructure as code capabilities, users can define and deploy logging infrastructure and configurations programmatically, ensuring consistency and repeatability across environments. Whether it's configuring logging with CloudWatch Logs, defining log streams and retention policies, or integrating with other logging and analysis tools such as Elasticsearch and Fluentd, Terraform provides a flexible and powerful platform for implementing centralized logging solutions that meet the needs of modern cloud-based environments.

BOOK 3
OPTIMIZING AWS INFRASTRUCTURE
ADVANCED TERRAFORM STRATEGIES

ROB BOTWRIGHT

Chapter 1: Infrastructure Cost Optimization Techniques

Resource right-sizing strategies are essential for optimizing cloud infrastructure costs while ensuring optimal performance and resource utilization. With the dynamic and scalable nature of cloud environments, it's crucial for organizations to continuously assess and adjust the size of their cloud resources to match workload requirements and avoid over-provisioning or under-provisioning. Next, we will explore various techniques and best practices for right-sizing cloud resources using Terraform, a popular infrastructure as code (IaC) tool, along with other complementary technologies such as AWS Cost Explorer and CloudWatch metrics.

The first step in implementing resource right-sizing strategies is to gather data on resource utilization and performance metrics to identify opportunities for optimization. AWS provides tools such as AWS Cost Explorer and CloudWatch metrics to analyze resource usage and performance metrics, enabling organizations to identify underutilized or overutilized resources that can be right-sized. For example, users can use AWS Cost Explorer to analyze historical usage data and identify cost-saving opportunities, or CloudWatch metrics to monitor CPU, memory, and storage utilization of EC2 instances.

Once potential optimization opportunities are identified, Terraform can be used to adjust the size of cloud resources to better match workload requirements. For example, users can define Terraform configurations to resize EC2 instances or change instance types based on performance metrics and workload characteristics. The following Terraform configuration demonstrates how to resize an EC2 instance to a smaller instance type:

hclCopy code

```
# Define EC2 instance resource "aws_instance" "example_instance" { instance_type = "t2.micro" # Other instance configuration... }
```

In this configuration, the instance type of the EC2 instance is changed to "t2.micro," which has lower resource specifications compared to the previous instance type, allowing users to reduce costs while maintaining sufficient performance for the workload.

Furthermore, Terraform enables users to automate the process of right-sizing resources by implementing dynamic scaling policies based on workload demand. For example, users can define Terraform configurations to automatically adjust the size of an Auto Scaling group based on CPU utilization or other performance metrics. The following Terraform configuration demonstrates how to define an Auto Scaling group with scaling policies based on CPU utilization:

hclCopy code

```
# Define Auto Scaling group resource "aws_autoscaling_group" "example_asg" { # Auto
```

Scaling group configuration... min_size = 1 max_size = 10 scaling_policy { name = "cpu-scaling-policy" adjustment_type = "ChangeInCapacity" scaling_adjustment = 1 cooldown = 300 metric_aggregation_type = "Average" target_tracking_configuration { predefined_metric_specification { predefined_metric_type = "ASGAverageCPUUtilization" }}}}

In this configuration, an Auto Scaling group named "example_asg" is defined with a scaling policy based on CPU utilization. When the average CPU utilization exceeds a predefined threshold, the Auto Scaling group will automatically scale out by adding more instances to handle the increased workload, ensuring optimal performance and resource utilization.

Moreover, Terraform can be integrated with third-party monitoring and optimization tools to enhance right-sizing strategies. For example, users can integrate Terraform with tools such as CloudHealth or Spot by NetApp to gain deeper insights into resource utilization and cost patterns and automate optimization actions based on predefined policies and rules.

In summary, resource right-sizing strategies are essential for optimizing cloud infrastructure costs and performance in dynamic and scalable environments. By leveraging Terraform's infrastructure as code capabilities, organizations can automate the process of adjusting the size of cloud resources based on workload demand and performance metrics, ensuring optimal resource utilization and cost efficiency. Whether it's resizing EC2 instances, implementing dynamic scaling policies, or integrating with third-party optimization tools, Terraform provides a flexible and powerful platform for implementing resource right-sizing strategies that meet the needs of modern cloud-based environments.

Leveraging AWS Cost Explorer and Cost Allocation Tags is pivotal for gaining deeper insights into cloud spending patterns and effectively managing costs in AWS environments. AWS Cost Explorer offers a comprehensive suite of tools and features that enable users to analyze historical spending data, forecast future costs, and identify cost-saving opportunities. Additionally, Cost Allocation Tags provide granular visibility into resource utilization and cost attribution, allowing organizations to allocate costs accurately and optimize resource allocation. Next, we will delve into how to leverage AWS Cost Explorer and Cost Allocation Tags effectively using CLI commands and Terraform configurations to optimize cloud spending and maximize cost efficiency.

To begin leveraging AWS Cost Explorer, users can utilize the AWS Command Line Interface (CLI) to access cost and usage data programmatically. The **aws ce get-cost-and-usage** command allows users to retrieve detailed cost and usage data for specific time periods, services, and cost dimensions. For instance, to retrieve cost and usage data for a specific month, users can execute the following CLI command:

bashCopy code

```
aws ce get-cost-and-usage --time-period Start=2024-01-
01,End=2024-01-31 --granularity MONTHLY
```

This command fetches cost and usage data for the month of January 2024, providing insights into spending patterns and cost drivers.

Furthermore, AWS Cost Explorer enables users to visualize cost data using various predefined and custom reports. The **aws ce get-cost-and-usage-with-resources** command allows users to generate a cost and usage report with resource-level granularity, providing detailed insights into resource utilization and cost allocation. For example, to generate a cost and usage report with resource-level granularity for the month of January 2024, users can execute the following CLI command:

bashCopy code

```
aws ce get-cost-and-usage-with-resources --time-
period Start=2024-01-01,End=2024-01-31 --granularity
MONTHLY --metrics "BlendedCost" --group-by
Type=DIMENSION,Key=SERVICE
```

This command generates a cost and usage report with resource-level granularity, grouped by AWS service, providing insights into the cost distribution across different services.

In addition to AWS Cost Explorer, Cost Allocation Tags play a crucial role in cost management and optimization by providing granular visibility into resource utilization and cost attribution. By tagging AWS resources with custom metadata, organizations can categorize and track resource usage based on different attributes such as environment, department, or project. This enables accurate cost allocation and chargeback, facilitating cost accountability and optimization efforts.

To leverage Cost Allocation Tags effectively, users can define tag policies and enforce tagging standards using Terraform configurations. The following Terraform configuration demonstrates how to define a tag policy to enforce tagging standards for EC2 instances:

hclCopy code

```
# Define tag policy resource "aws_resourcegroups_tagging_policy" "example_tag_policy" { policy = jsonencode({ "tags" = { "Environment" = "Development", "Department" = "Engineering" }, "compliance_rules" = [ { "tag_key" = "Environment", "compliance_type" = "MUST_EXIST" }, { "tag_key" = "Department", "compliance_type" = "MUST_EXIST" } ] }) }
```

In this configuration, a tag policy is defined to enforce the presence of "Environment" and "Department" tags on EC2 instances, ensuring consistent tagging standards and accurate cost allocation.

Moreover, Terraform enables users to automate the process of tagging AWS resources based on predefined rules and conditions. The following Terraform configuration demonstrates how to automatically tag EC2 instances based on specific criteria:

hclCopy code

```
# Automatically tag EC2 instances resource "aws_instance" "example_instance" { # Instance
```

configuration... tags = { "Environment" = "Production", "Department" = "Operations" } }

In this configuration, EC2 instances are automatically tagged with "Environment" set to "Production" and "Department" set to "Operations" during provisioning, ensuring consistent tagging and accurate cost attribution.

In summary, leveraging AWS Cost Explorer and Cost Allocation Tags is essential for gaining visibility into cloud spending patterns and optimizing costs in AWS environments. By utilizing CLI commands and Terraform configurations, organizations can analyze cost and usage data, generate custom reports, enforce tagging standards, and automate tagging processes, enabling effective cost management and optimization efforts. Whether it's analyzing cost data with AWS Cost Explorer, defining tag policies with Terraform, or automatically tagging resources based on predefined rules, AWS Cost Explorer and Cost Allocation Tags provide powerful tools for optimizing cloud spending and maximizing cost efficiency.

Chapter 2: Performance Optimization with Terraform and AWS

Implementing AWS Performance Monitoring Tools is crucial for maintaining optimal performance and availability of cloud-based applications and infrastructure. AWS offers a range of monitoring tools and services that enable organizations to monitor various aspects of their AWS environments, including resource utilization, application performance, and infrastructure health. Next, we will explore how to implement AWS Performance Monitoring Tools effectively using CLI commands and configurations to ensure proactive monitoring and timely detection of performance issues.

One of the fundamental tools for performance monitoring in AWS is Amazon CloudWatch, a monitoring and observability service that provides real-time insights into AWS resources and applications. CloudWatch enables users to collect and track metrics, monitor logs, set alarms, and visualize performance data through dashboards. To begin implementing performance monitoring with CloudWatch, users can utilize the AWS CLI to create CloudWatch alarms for monitoring key metrics. The **aws cloudwatch put-metric-alarm** command allows users to create alarms for specific metrics, such as CPU utilization, memory utilization, or network traffic.

For example, to create an alarm for high CPU utilization on an EC2 instance, users can execute the following CLI command:

bashCopy code

```
aws cloudwatch put-metric-alarm --alarm-name "HighCPUUtilization" --metric-name "CPUUtilization" --namespace "AWS/EC2" --statistic "Average" --period 300 --threshold 80 --comparison-operator "GreaterThanOrEqualToThreshold" --evaluation-periods 3 --alarm-actions "arn:aws:sns:us-east-1:123456789012:my-topic"
```

This command creates an alarm named "HighCPUUtilization" that triggers when the average CPU utilization of the specified EC2 instance exceeds 80% for three consecutive evaluation periods.

In addition to CloudWatch alarms, CloudWatch Logs enables users to monitor and analyze log data from AWS resources and applications in real-time. Users can use the AWS CLI to configure log streams, define log group retention policies, and filter log data for analysis. For example, to create a log group and stream for an AWS Lambda function, users can execute the following CLI commands:

bashCopy code

```
aws logs create-log-group --log-group-name "/aws/lambda/my-lambda-function" aws logs create-log-stream --log-group-name "/aws/lambda/my-
```

lambda-function" --log-stream-name "my-lambda-function-stream"

These commands create a log group named "/aws/lambda/my-lambda-function" and a corresponding log stream named "my-lambda-function-stream" for logging Lambda function output.

Furthermore, AWS X-Ray is a distributed tracing service that enables users to trace requests as they travel through AWS services and microservices-based applications. X-Ray provides insights into request latency, error rates, and service dependencies, allowing users to identify performance bottlenecks and optimize application performance. To implement performance monitoring with X-Ray, users can instrument their applications with the X-Ray SDK or use AWS services that support X-Ray tracing. For example, to enable X-Ray tracing for an AWS Lambda function, users can execute the following CLI command:

bashCopy code

aws lambda update-function-configuration --function-name my-lambda-function --tracing-config Mode=Active

This command enables active tracing for the specified Lambda function, allowing X-Ray to trace incoming requests and capture performance data.

Moreover, AWS Application Insights is a monitoring and troubleshooting service that helps users detect and diagnose issues in AWS resources and applications. Application Insights provides automated insights, anomaly detection, and root cause analysis capabilities, enabling users to identify and resolve performance issues quickly. To set up Application Insights for an application, users can use the AWS Management Console or CLI to configure monitoring settings and enable insights. For example, to enable Application Insights for an Amazon RDS database, users can execute the following CLI command:

bashCopy code

```
aws rds add-monitoring-role-to-db-instance --db-instance-identifier my-rds-instance --monitoring-role-arn
arn:aws:iam::123456789012:role/ApplicationInsights-Role
```

This command adds the necessary IAM role to the specified RDS instance to enable monitoring and insights collection by Application Insights.

In summary, implementing AWS Performance Monitoring Tools is essential for maintaining optimal performance and availability of AWS resources and applications. By leveraging tools such as Amazon CloudWatch, AWS X-Ray, and AWS Application Insights, organizations can monitor key metrics, analyze performance data, and detect issues proactively to ensure a seamless user experience. Whether it's configuring CloudWatch alarms, tracing requests with X-Ray, or enabling insights with Application Insights, AWS provides a comprehensive suite of monitoring tools and services to meet the performance monitoring needs of modern cloud environments.

Optimizing resource configuration for performance is a critical aspect of managing cloud-based environments effectively. In AWS, optimizing resource configuration involves fine-tuning various settings and parameters to ensure optimal performance, scalability, and cost-efficiency. This chapter explores techniques and best practices for optimizing resource configuration in AWS environments using CLI commands and configuration options to enhance performance and maximize resource utilization.

One of the key aspects of optimizing resource configuration is selecting the appropriate instance types for EC2 instances. AWS offers a wide range of EC2 instance types optimized for different use cases, such as compute-intensive workloads, memory-intensive applications, or storage-intensive tasks. To select the optimal instance type for a specific workload, users can leverage the AWS CLI to retrieve information about available instance types and their specifications. The **aws ec2 describe-instance-types** command provides detailed information about instance types, including vCPU, memory, storage, and network performance. For example, to list all available instance types in the us-east-1 region, users can execute the following CLI command:

bashCopy code

aws ec2 describe-instance-types --region us-east-1

This command returns a list of instance types along with their specifications, allowing users to compare performance characteristics and select the most suitable instance type for their workload.

In addition to selecting the appropriate instance type, optimizing resource configuration involves configuring instance attributes such as CPU, memory, storage, and network settings to meet performance requirements. For example, users can adjust CPU and memory allocations for EC2 instances based on workload demands using CLI commands or Terraform configurations. The **aws ec2 modify-instance-attributes** command allows users to modify instance attributes such as CPU options, memory settings, and network configurations. For instance, to modify the CPU options for an EC2 instance to enable enhanced networking, users can execute the following CLI command:

bashCopy code

aws ec2 modify-instance-attribute --instance-id i-1234567890abcdef0 --ena-support

This command enables Enhanced Networking with the Elastic Network Adapter (ENA) for the specified EC2 instance, improving network performance and throughput.

Furthermore, optimizing resource configuration involves fine-tuning performance parameters and settings for AWS services such as Amazon RDS,

Amazon Redshift, and Amazon DynamoDB. For example, users can optimize database performance by configuring parameters such as storage type, instance size, IOPS, and read/write capacity for Amazon RDS and DynamoDB instances. The **aws rds modify-db-instance** command allows users to modify database instance settings such as instance class, storage type, and allocated storage. For instance, to modify the instance class and storage type for an RDS instance, users can execute the following CLI command:

bashCopy code

aws rds modify-db-instance --db-instance-identifier mydbinstance --db-instance-class db.t3.medium --storage-type gp2

This command modifies the instance class to db.t3.medium and the storage type to General Purpose SSD (gp2) for the specified RDS instance, enhancing performance and scalability.

Moreover, optimizing resource configuration involves implementing best practices for security, compliance, and reliability. Users can leverage AWS Identity and Access Management (IAM) policies, network security groups, and encryption options to enhance security and compliance posture. Additionally, users can implement redundancy and fault tolerance mechanisms such as multi-AZ deployments and data replication to improve reliability and availability.

In summary, optimizing resource configuration for performance is essential for maximizing the efficiency and effectiveness of AWS environments. By selecting the appropriate instance types, configuring performance parameters, and implementing best practices for security and reliability, organizations can ensure optimal performance, scalability, and cost-efficiency in their cloud deployments. Whether it's fine-tuning CPU and memory allocations, optimizing database settings, or enhancing security and compliance posture, optimizing resource configuration plays a crucial role in achieving peak performance and maximizing the value of AWS resources.

Chapter 3: High Availability and Disaster Recovery Planning

Designing High Availability (HA) architectures with Terraform is a crucial aspect of building resilient and fault-tolerant cloud infrastructures in AWS. High Availability architectures are designed to minimize downtime and ensure continuous operation of applications and services by distributing workloads across multiple availability zones (AZs) and implementing redundancy and failover mechanisms. Next, we will explore techniques and best practices for designing HA architectures using Terraform, including deploying multi-AZ deployments, implementing auto-scaling and load balancing, and configuring fault-tolerant storage solutions.

One of the fundamental principles of designing HA architectures is distributing workloads across multiple availability zones to mitigate the risk of single points of failure. AWS offers multiple availability zones within each region, allowing users to deploy resources across separate data centers for redundancy and resilience. Terraform enables users to define multi-AZ architectures by specifying multiple availability zones in their infrastructure configurations. For example, when provisioning EC2 instances with Terraform, users can specify multiple subnets in different availability zones to ensure high availability. The

availability_zones parameter in Terraform allows users to specify a list of availability zones for deploying resources. Here's an example Terraform configuration snippet for provisioning EC2 instances across multiple availability zones:

hclCopy code

```
resource "aws_instance" "example" { count = 3 ami = "ami-12345678" instance_type = "t2.micro" subnet_id = element(var.subnets, count.index) availability_zone = element(data.aws_availability_zones.available.names, count.index) }
```

This Terraform configuration creates three EC2 instances and distributes them across three different subnets in separate availability zones, ensuring high availability and fault tolerance.

Furthermore, implementing auto-scaling and load balancing is essential for ensuring high availability and performance scalability in HA architectures. AWS provides auto-scaling groups and load balancing services such as Elastic Load Balancing (ELB) to automatically adjust capacity and distribute incoming traffic across multiple instances. Terraform enables users to define auto-scaling groups and load balancers in their infrastructure configurations to achieve HA and scalability. For example, users can use Terraform to create an auto-scaling group with launch configurations and attach it to an Application Load Balancer (ALB) to distribute traffic across instances dynamically. Here's an example Terraform configuration for creating an auto-scaling group and ALB:

```hcl
hclCopy code
resource "aws_autoscaling_group" "example" {
desired_capacity = 3 min_size = 2 max_size = 5
launch_configuration =
aws_launch_configuration.example.name
vpc_zone_identifier = var.subnets } resource
"aws_lb" "example" { name = "example-lb" internal =
false load_balancer_type = "application" subnets =
var.subnets } resource "aws_lb_target_group"
"example" { name = "example-target-group" port =
80 protocol = "HTTP" vpc_id = var.vpc_id depends_on
= [aws_lb.example] } resource "aws_lb_listener"
"example" { load_balancer_arn = aws_lb.example.arn
port = "80" protocol = "HTTP" default_action { type =
"forward" target_group_arn =
aws_lb_target_group.example.arn } }
```

This Terraform configuration creates an auto-scaling group with three instances and attaches it to an ALB to distribute incoming HTTP traffic across instances, ensuring high availability and scalability.

Additionally, configuring fault-tolerant storage solutions is crucial for ensuring data durability and availability in HA architectures. AWS offers storage services such as Amazon S3 and Amazon EBS that provide built-in redundancy and fault tolerance.

Terraform enables users to define resilient storage configurations by specifying replication options and redundancy settings. For example, when provisioning an Amazon S3 bucket with Terraform, users can specify cross-region replication to replicate data across multiple AWS regions for enhanced durability and availability. Here's an example Terraform configuration for creating an S3 bucket with cross-region replication:

hclCopy code

```
resource "aws_s3_bucket" "example" { bucket = "example-bucket" acl = "private" replication_configuration { role = aws_iam_role.replication_role.arn rules { id = "replicate-to-other-region" status = "Enabled" destination { bucket = "arn:aws:s3:::destination-bucket" storage_class = "STANDARD" } } } }
```

This Terraform configuration creates an S3 bucket named "example-bucket" with cross-region replication enabled, ensuring data redundancy and availability across multiple AWS regions.

In summary, designing High Availability architectures with Terraform is essential for building resilient and fault-tolerant cloud infrastructures in AWS. By leveraging Terraform's infrastructure-as-code capabilities, users can define multi-AZ deployments, implement auto-scaling and load balancing, and configure fault-tolerant storage solutions to achieve high availability and ensure continuous operation of applications and services. Whether it's distributing workloads across multiple availability zones, scaling resources dynamically, or replicating data for redundancy, Terraform provides the tools and capabilities to design and deploy highly available architectures in AWS effectively.

Implementing Disaster Recovery Strategies on AWS is crucial for ensuring business continuity and mitigating the impact of potential disasters or outages on critical applications and data. AWS provides a range of services and features that enable organizations to implement robust disaster recovery (DR) solutions to protect against various failure scenarios, including natural disasters, hardware failures, and human errors. Next, we will explore techniques and best practices for implementing disaster recovery strategies on AWS using CLI commands and configuration options to minimize downtime and data loss in the event of a disaster.

One of the fundamental aspects of implementing disaster recovery on AWS is designing a multi-region architecture to distribute workloads and data across geographically dispersed AWS regions. By deploying resources and applications across multiple regions, organizations can mitigate the impact of regional failures and ensure high availability and resilience. To deploy resources in multiple regions, users can leverage Terraform to define infrastructure as code (IaC) templates and provision resources in different regions simultaneously. The **terraform apply** command allows users to apply Terraform configurations and deploy resources across multiple regions. For example, to deploy an Amazon RDS database in two different regions, users can execute the following Terraform commands:

bashCopy code

terraform init terraform plan -var="region=us-east-1" terraform apply -var="region=us-east-1" terraform plan -var="region=us-west-2" terraform apply -var="region=us-west-2"

These commands initialize Terraform, generate an execution plan, and apply the Terraform configuration to provision an RDS database in the specified regions.

Furthermore, implementing disaster recovery strategies involves replicating data and resources across regions to ensure data consistency and redundancy. AWS offers services such as Amazon S3 cross-region replication and Amazon Aurora Global Database to replicate data across regions automatically. Users can configure cross-region replication using CLI commands or AWS Management Console to replicate objects from a source bucket to a destination bucket in a different region. For example, to enable cross-region replication for an S3 bucket, users can execute the following CLI commands:

bashCopy code

```
aws s3api put-bucket-replication --bucket source-
bucket --replication-configuration file://replication-
config.json
```
Where **replication-config.json** contains the replication configuration specifying the destination bucket and replication rules.

Moreover, implementing disaster recovery strategies involves automating failover and recovery processes to minimize downtime and ensure rapid recovery in the event of a disaster. AWS provides services such as Amazon Route 53 DNS failover, AWS Lambda, and Amazon CloudWatch Alarms to automate failover and recovery operations based on predefined criteria. Users can configure DNS failover using Route 53 health checks and failover policies to route traffic to healthy endpoints during an outage. For example, to configure DNS failover for a web application, users can create Route 53 health checks and associate them with failover records using CLI commands or AWS Management Console.

```
bashCopy code
aws route53 create-health-check --caller-reference
my-health-check --health-check-config file://health-
check-config.json
```

Where **health-check-config.json** contains the health check configuration specifying the endpoint and health check settings.

In addition to DNS failover, users can leverage AWS Lambda functions and CloudWatch Alarms to automate recovery actions such as restarting failed instances, restoring backups, and scaling resources. By combining Lambda functions with CloudWatch Alarms, users can define custom recovery workflows and trigger automated actions based on predefined thresholds or conditions.

In summary, implementing disaster recovery strategies on AWS is essential for ensuring business continuity and resilience in the face of unforeseen events. By designing multi-region architectures, replicating data across regions, and automating failover and recovery processes, organizations can minimize downtime, mitigate data loss, and maintain uninterrupted service availability during disasters. Whether it's deploying resources across multiple regions with Terraform, configuring cross-region replication with S3, or automating failover with Route 53 DNS, AWS offers a comprehensive set of tools and services to help organizations build robust and reliable disaster recovery solutions.

Chapter 4: Advanced Load Balancing and Traffic Management

Advanced Load Balancer Configuration with Terraform is essential for optimizing the performance, scalability, and resilience of applications deployed on AWS. Load balancers play a crucial role in distributing incoming traffic across multiple instances or targets to ensure high availability and fault tolerance. Next, we will explore advanced techniques and best practices for configuring and managing load balancers using Terraform, along with CLI commands to deploy these configurations effectively.

One of the key aspects of advanced load balancer configuration is fine-tuning health checks to improve the reliability and responsiveness of load balancers. Health checks allow load balancers to monitor the health and status of registered instances or targets and route traffic only to healthy ones. Terraform provides options to configure health checks with custom settings such as interval, timeout, and thresholds to suit specific application requirements. Users can define health check configurations within their Terraform scripts using the **aws_lb_target_group** resource. For example, to create a target group with custom health check settings, users can use the following Terraform configuration:

hclCopy code

resource "aws_lb_target_group" "example" { name = "example-target-group" port = 80 protocol = "HTTP" vpc_id = aws_vpc.example.id health_check { path = "/health" port = "traffic-port" protocol = "HTTP" timeout = 5 interval = 30 healthy_threshold = 2 unhealthy_threshold = 2 } }

This Terraform configuration creates an Application Load Balancer target group with a custom HTTP health check configuration.

Additionally, implementing advanced load balancer configurations involves optimizing routing algorithms and rules to distribute traffic effectively based on various criteria such as request attributes, URL paths, or session affinity. Terraform allows users to define routing rules and conditions using the **aws_lb_listener_rule** resource. For example, to configure a listener rule to route traffic based on the request path, users can define the following Terraform configuration:

hclCopy code

```
resource "aws_lb_listener_rule" "example" {
listener_arn = aws_lb_listener.example.arn priority =
100 action { type = "forward" target_group_arn =
aws_lb_target_group.example.arn } condition {
host_header { values = ["example.com"] } path_pattern
{ values = ["/api/*"] } } }
```

This Terraform configuration creates a listener rule that forwards requests with the specified host header and path pattern to the designated target group.

Moreover, advanced load balancer configuration involves integrating with other AWS services and features such as AWS WAF (Web Application Firewall), AWS Shield, and AWS Lambda to enhance security, mitigate DDoS attacks, and implement custom request processing logic. Terraform provides resources and modules to configure these integrations seamlessly within infrastructure as code (IaC) templates. For example, to associate an AWS WAF web ACL with a load balancer, users can define the following Terraform configuration:

hclCopy code

```
resource "aws_wafv2_web_acl_association" "example"
{ resource_arn = aws_lb.example.arn web_acl_arn =
aws_wafv2_web_acl.example.arn }
```

This Terraform configuration associates an AWS WAFv2 web ACL with the specified load balancer to protect against common web-based attacks.

In summary, advanced load balancer configuration with Terraform enables organizations to optimize the performance, scalability, and security of their applications deployed on AWS. By fine-tuning health checks, optimizing routing rules, and integrating with other AWS services, organizations can build resilient and high-performance architectures that meet the demands of modern applications. Whether it's customizing health check settings, defining sophisticated routing rules, or implementing security controls with AWS WAF, Terraform provides a flexible and powerful platform for managing advanced load balancer configurations efficiently and effectively.

Chapter 5: Security Best Practices for AWS Infrastructure

Implementing Security Controls with Terraform is imperative for safeguarding cloud environments and ensuring compliance with industry regulations and best practices. Security is a top priority for organizations deploying applications and infrastructure on cloud platforms like AWS, and Terraform provides powerful capabilities to define, configure, and manage security controls as code. Next, we will delve into various security controls that can be implemented using Terraform, along with CLI commands and deployment techniques to enforce security policies effectively.

One of the fundamental aspects of implementing security controls with Terraform is managing access and permissions using Identity and Access Management (IAM). IAM allows organizations to control who can access AWS resources and what actions they can perform. Terraform enables users to define IAM policies, roles, and permissions as code using the **aws_iam_policy** and **aws_iam_role** resources. For example, to create an IAM policy that grants read-only access to S3 buckets, users can define the following Terraform configuration:
hclCopy code

```
resource "aws_iam_policy" "s3_read_only_policy" {
name = "s3-read-only-policy" description = "Allows
read-only access to S3 buckets" policy = <<EOF {
"Version": "2012-10-17", "Statement": [ { "Effect":
"Allow", "Action": [ "s3:GetObject", "s3:ListBucket" ],
"Resource":    [    "arn:aws:s3:::example-bucket",
"arn:aws:s3:::example-bucket/*" ] } ] } EOF }
```
This Terraform configuration creates an IAM policy
that allows the specified actions (**s3:GetObject** and
s3:ListBucket) on the specified S3 bucket (**example-
bucket**).

Additionally, implementing security controls involves
configuring network security using Virtual Private
Cloud (VPC) security groups and Network Access
Control Lists (NACLs). VPC security groups act as
virtual firewalls to control inbound and outbound
traffic to EC2 instances, while NACLs provide subnet-
level security controls. Terraform allows users to
define security group rules and NACL rules as code
using the **aws_security_group_rule** and
aws_network_acl_rule resources. For example, to
create a security group rule that allows inbound SSH
traffic from a specific IP range, users can define the
following Terraform configuration:

hclCopy code
```
resource "aws_security_group_rule" "ssh_ingress" {
type = "ingress" from_port = 22 to_port = 22 protocol
```

= "tcp" cidr_blocks = ["10.0.0.0/24"] security_group_id = aws_security_group.example.id } This Terraform configuration creates an ingress rule in the specified security group (**aws_security_group.example**) that allows inbound SSH traffic from the specified CIDR block (**10.0.0.0/24**).

Moreover, implementing security controls with Terraform involves enforcing encryption and data protection policies across various AWS services. Terraform provides resources and modules to configure encryption settings for services such as S3, EBS volumes, RDS databases, and more. For example, to enable server-side encryption for an S3 bucket, users can define the following Terraform configuration:

hclCopy code

```
resource "aws_s3_bucket" "example" { bucket = "example-bucket" acl = "private" server_side_encryption_configuration { rule { apply_server_side_encryption_by_default { sse_algorithm = "AES256" } } } }
```

This Terraform configuration creates an S3 bucket with server-side encryption enabled using AES256 encryption algorithm.

Furthermore, implementing security controls involves continuous monitoring and auditing of AWS resources to detect and respond to security threats and vulnerabilities. Terraform integrates with AWS services such as CloudTrail, CloudWatch, and Config to enable automated monitoring and logging of infrastructure changes and security events. Users can configure CloudTrail trails, CloudWatch alarms, and Config rules using Terraform resources and modules to monitor and alert on security-related activities. For example, to create a CloudTrail trail that logs all API activity in an AWS account, users can define the following Terraform configuration:

hclCopy code

```
resource "aws_cloudtrail" "example" { name = "example-trail" s3_bucket_name = aws_s3_bucket.example_bucket.id enable_logging = true include_global_service_events = true }
```

This Terraform configuration creates a CloudTrail trail that logs API activity to the specified S3 bucket.

In summary, implementing security controls with Terraform is essential for protecting cloud environments and ensuring compliance with security standards and regulations. Whether it's managing access permissions with IAM, configuring network security with VPC security groups and NACLs, enforcing encryption policies, or monitoring and auditing infrastructure with CloudTrail and CloudWatch, Terraform provides a comprehensive set of tools and capabilities to help organizations build secure and resilient AWS deployments. By treating security as code and using Terraform to define and manage security controls alongside infrastructure, organizations can achieve greater consistency, visibility, and automation in their security practices.

Compliance auditing and security monitoring are critical components of maintaining the integrity and security of cloud environments. With the increasing complexity of regulatory requirements and the evolving threat landscape, organizations must implement robust processes for auditing and monitoring their infrastructure to ensure compliance with industry standards and regulations and to detect and respond to security incidents effectively.

In AWS, compliance auditing involves assessing infrastructure configurations against established security benchmarks and regulatory frameworks such as HIPAA, GDPR, PCI DSS, and more. AWS provides tools and services to facilitate compliance auditing, including AWS Config, AWS Security Hub, and AWS Audit Manager. These services enable organizations to continuously monitor their AWS resources for compliance violations and security risks and to generate detailed reports for auditing purposes.

To initiate compliance auditing in AWS using AWS Config, users can enable AWS Config recording for their AWS account using the AWS Management Console or the AWS CLI. The following CLI command enables AWS Config recording:

bashCopy code

```
aws configservice start-configuration-recorder --configuration-recorder-name default
```

Once AWS Config recording is enabled, AWS Config will automatically record configuration changes to AWS resources and evaluate resource configurations against predefined rulesets called Config rules. Users can create custom Config rules or use managed Config rules provided by AWS to check for compliance with specific security standards.

For example, to enable a managed Config rule that checks for unrestricted SSH access to EC2 instances, users can use the following CLI command:

```
bashCopy code
aws configservice put-config-rule --config-rule
'{"Type":"AWS::Config::ConfigRule","Properties":{"ConfigRuleName":"ec2-security-group-ssh","Scope":{"ComplianceResourceTypes":["AWS::EC2::SecurityGroup"]},"Source":{"Owner":"AWS","SourceIdentifier":"SECURITY_GROUP_OPEN_PORTS_CHECK"}}'
```

This command creates a new Config rule named **ec2-security-group-ssh** that checks for security groups allowing unrestricted SSH access.

In addition to AWS Config, organizations can leverage AWS Security Hub to aggregate and prioritize security findings from multiple AWS services and third-party security tools. Security Hub provides a centralized dashboard for monitoring security posture and compliance status across AWS accounts, as well as automated compliance checks and remediation recommendations.

To enable AWS Security Hub in an AWS account, users can use the following CLI command:
bashCopy code

```
aws securityhub enable-security-hub
```

Once Security Hub is enabled, it will automatically ingest security findings from integrated AWS services such as Amazon GuardDuty, Amazon Inspector, AWS Config, and more, allowing organizations to gain real-time visibility into security threats and vulnerabilities.

Furthermore, AWS Audit Manager simplifies the process of conducting audit assessments and managing compliance requirements by providing prebuilt frameworks and automated workflows for assessing AWS resources against regulatory standards. Audit Manager enables organizations to define assessment scopes, select predefined frameworks or create custom controls, and automate evidence collection and reporting.

To initiate an audit assessment in AWS Audit Manager, users can use the following CLI command:
bashCopy code

```
aws auditmanager create-assessment --name "HIPAA Compliance Assessment" --framework-id "aws/hitrustcsf_v9_2_2020" --assessment-reports-destination '{"destinationType": "S3","destination":"arn:aws:s3:::audit-reports-bucket"}'
```

This command creates a new audit assessment named "HIPAA Compliance Assessment" based on the

HITRUST CSF framework and specifies an S3 bucket as the destination for assessment reports.

In summary, compliance auditing and security monitoring are essential practices for ensuring the security and regulatory compliance of AWS environments. By leveraging AWS services such as AWS Config, AWS Security Hub, and AWS Audit Manager, organizations can automate the process of auditing infrastructure configurations, monitoring for security threats, and demonstrating compliance with industry standards and regulations. With the ability to define, deploy, and manage compliance controls as code using the AWS CLI and APIs, organizations can achieve greater agility, scalability, and security in their cloud environments.

Chapter 6: Advanced IAM Policies and Role Management

Implementing least privilege access controls is a crucial aspect of maintaining the security and integrity of cloud environments. By granting users and resources only the permissions they need to perform their tasks, organizations can minimize the risk of unauthorized access and potential security breaches. Terraform, with its infrastructure as code approach, provides a powerful means to implement and manage least privilege access controls in AWS environments.

To begin implementing least privilege access controls with Terraform, organizations should first define their IAM (Identity and Access Management) policies and roles. IAM policies specify the permissions granted to users, groups, or roles, while IAM roles define the set of permissions that an AWS service or resource can assume. By carefully crafting IAM policies and roles, organizations can enforce the principle of least privilege and restrict access to only the necessary actions and resources.

For example, to create an IAM policy that grants read-only access to S3 buckets, organizations can define the following Terraform configuration:

hclCopy code

resource "aws_iam_policy" "s3_read_only" { name = "s3-read-only-policy" description = "Provides read-

199

only access to S3 buckets" policy = <<EOF { "Version": "2012-10-17", "Statement": [{ "Effect": "Allow", "Action": "s3:Get*", "Resource": "arn:aws:s3:::*" }] } EOF }

This Terraform configuration creates an IAM policy named "s3-read-only-policy" that allows users or roles attached to this policy to perform read-only actions on all S3 buckets in the AWS account.

Next, organizations can create IAM roles and attach the defined IAM policies to these roles using Terraform. For example, to create an IAM role named "s3-read-only-role" and attach the previously defined "s3-read-only-policy" to it, organizations can use the following Terraform configuration:

```
hclCopy code
resource "aws_iam_role" "s3_read_only_role" {
  name = "s3-read-only-role" assume_role_policy =
jsonencode({ "Version": "2012-10-17", "Statement": [
  { "Effect": "Allow", "Principal": { "Service":
"ec2.amazonaws.com" }, "Action": "sts:AssumeRole"
} ] }) } resource "aws_iam_role_policy_attachment"
"s3_read_only_attachment" { role =
aws_iam_role.s3_read_only_role.name policy_arn =
aws_iam_policy.s3_read_only.arn }
```

This Terraform configuration creates an IAM role named "s3-read-only-role" and attaches the "s3-read-only-policy" to it. The assume_role_policy specifies that EC2 instances can assume this role.

Once the IAM policies and roles are defined and configured, organizations can provision resources such as EC2 instances, Lambda functions, or ECS tasks with the necessary permissions by associating them with the appropriate IAM roles. For example, to launch an EC2 instance with the "s3-read-only-role" IAM role, organizations can use the following Terraform configuration:

hclCopy code

```
resource "aws_instance" "example_instance" { ami = "ami-0c55b159cbfafe1f0" instance_type = "t2.micro" iam_instance_profile = aws_iam_instance_profile.example_profile.name }
```

This Terraform configuration provisions an EC2 instance and associates it with the IAM role "s3-read-only-role" using the iam_instance_profile attribute.

By leveraging Terraform to define and manage IAM policies, roles, and resource permissions, organizations can implement least privilege access controls effectively and ensure that their AWS environments adhere to security best practices. Through infrastructure as code principles, changes to access controls can be tracked, versioned, and audited, providing greater visibility and control over security configurations. Additionally, Terraform's declarative syntax allows organizations to enforce consistent access control policies across different environments and manage complex IAM configurations with ease.

Role-Based Access Control (RBAC) strategies play a pivotal role in managing access to resources within AWS environments, ensuring that users and services are granted permissions based on their roles and responsibilities. RBAC allows organizations to implement granular access controls, thereby reducing the risk of unauthorized access and ensuring compliance with security requirements. In AWS, RBAC is typically implemented using AWS Identity and Access Management (IAM), which provides a comprehensive set of tools for defining and managing roles, policies, and permissions.

To begin implementing RBAC strategies on AWS, organizations need to define the roles that users and services will assume and the corresponding permissions associated with each role. This involves creating IAM policies that specify the actions and resources that users or services are allowed to access. These policies can be attached to IAM roles, groups, or directly to users, depending on the desired access control model.

For example, organizations can create an IAM policy that grants read-only access to S3 buckets:

```
bashCopy code
aws iam create-policy --policy-name s3-read-only-policy --policy-document file://s3-read-only-policy.json
```

In this command, **aws iam create-policy** is used to create a new IAM policy named "s3-read-only-policy", and the policy document is specified using the **--policy-document** parameter, referencing a JSON file (**s3-read-only-policy.json**) containing the policy definition.

Once the policy is created, it can be attached to IAM roles or users. Organizations can create IAM roles representing different roles or responsibilities within their organization, such as "developer", "administrator", or "auditor". These roles can then be associated with the appropriate IAM policies to grant the necessary permissions.

bashCopy code

aws iam create-role --role-name developer-role --assume-role-policy-document file://trust-policy.json

In this example, **aws iam create-role** is used to create a new IAM role named "developer-role", and the trust policy document is specified using the **--assume-role-policy-document** parameter, referencing a JSON file (**trust-policy.json**) containing the trust relationship definition, which specifies the entities allowed to assume the role.

Once the IAM role is created, the previously defined policy can be attached to it:

bashCopy code

aws iam attach-role-policy --role-name developer-role   --policy-arn

arn:aws:iam::123456789012:policy/s3-read-only-policy

In this command, **aws iam attach-role-policy** is used to attach the "s3-read-only-policy" to the "developer-role". The **--policy-arn** parameter specifies the ARN (Amazon Resource Name) of the IAM policy to attach.

By following this approach, organizations can effectively implement RBAC strategies on AWS, ensuring that users and services have the appropriate level of access to resources based on their roles and responsibilities. This helps enforce the principle of least privilege, mitigating the risk of unauthorized access and potential security breaches. Additionally, RBAC simplifies access management and auditing processes, allowing organizations to maintain a secure and compliant AWS environment.

Chapter 7: Infrastructure Monitoring and Alerting Strategies

Implementing monitoring solutions with Terraform is crucial for maintaining the health, performance, and security of infrastructure and applications deployed on AWS. Monitoring enables organizations to proactively identify and address issues, optimize resource utilization, and ensure compliance with service level agreements (SLAs) and regulatory requirements. Terraform provides a powerful framework for automating the deployment and configuration of monitoring tools, allowing organizations to define monitoring infrastructure as code and manage it alongside their application infrastructure.

One of the key components of monitoring solutions is collecting and aggregating metrics and logs from various AWS services and resources. AWS provides services like Amazon CloudWatch and AWS CloudTrail for monitoring and logging, which can be integrated with Terraform to automate their deployment and configuration.

For example, organizations can use Terraform to create a CloudWatch dashboard that displays key metrics and alarms for monitoring EC2 instances: bashCopy code

```
resource                "aws_cloudwatch_dashboard"
"example_dashboard" { dashboard_body = <<EOF {
"widgets": [ { "type": "metric", "x": 0, "y": 0, "width":
12, "height": 6, "properties": { "metrics": [ [
"AWS/EC2",     "CPUUtilization",     "InstanceId",
"${aws_instance.example.id}" ] ], "title": "EC2 CPU
Utilization", "period": 300, "stat": "Average",
"region": "us-west-2" } } ] } EOF }
```

In this Terraform configuration, an **aws_cloudwatch_dashboard** resource is defined to create a CloudWatch dashboard. The **dashboard_body** attribute specifies the JSON representation of the dashboard, including the metrics to be displayed (in this case, CPU utilization of an EC2 instance) and their properties.

Additionally, Terraform can be used to deploy and configure monitoring agents or agents-based solutions on EC2 instances to collect additional metrics and logs. For example, organizations can use the AWS Systems Manager (SSM) Agent to collect system metrics and logs from EC2 instances:

bashCopy code

```
resource "aws_instance" "example" { ami = "ami-
12345678" instance_type = "t2.micro" tags = { Name
= "example-instance" } user_data = <<-EOF
#!/bin/bash sudo yum install -y amazon-ssm-agent
sudo systemctl start amazon-ssm-agent EOF }
```

In this Terraform configuration, an **aws_instance** resource is defined to create an EC2 instance, and the

user_data attribute is used to specify a script that installs and starts the SSM Agent on the instance.

Furthermore, Terraform can be leveraged to configure CloudWatch alarms and notifications to alert operators when certain thresholds are exceeded or anomalies are detected in the monitored metrics.

bashCopy code

```
resource "aws_cloudwatch_metric_alarm" "example_alarm" { alarm_name = "example-alarm" comparison_operator = "GreaterThanOrEqualToThreshold" evaluation_periods = 2 metric_name = "CPUUtilization" namespace = "AWS/EC2" period = 120 statistic = "Average" threshold = 90 alarm_description = "This metric monitors CPU utilization" alarm_actions = [aws_sns_topic.example_topic.arn] }
```

In this example, an **aws_cloudwatch_metric_alarm** resource is defined to create a CloudWatch alarm that monitors CPU utilization of EC2 instances. The **threshold** attribute specifies the CPU utilization threshold (90%), and the **alarm_actions** attribute specifies the Amazon Resource Name (ARN) of the SNS topic to which the alarm sends notifications.

By utilizing Terraform to implement monitoring solutions on AWS, organizations can automate the deployment, configuration, and management of monitoring infrastructure, ensuring visibility and control over their cloud environments. This approach enables organizations to maintain the reliability, performance, and security of their applications and infrastructure, ultimately enhancing the overall operational efficiency and resilience of their AWS deployments.

Designing effective alerting policies for AWS infrastructure is crucial for maintaining the health, performance, and security of cloud environments. Alerting policies help organizations identify and respond to potential issues or anomalies in real-time, allowing them to mitigate risks and minimize service disruptions. In AWS, alerting policies are typically implemented using Amazon CloudWatch, a monitoring and observability service that provides metrics, logs, and alarms for AWS resources and applications.

To design alerting policies, organizations first need to identify the key metrics and thresholds that are relevant to their infrastructure and applications. These metrics may include CPU utilization, memory usage, disk I/O, network traffic, error rates, latency, and more. By defining appropriate thresholds for these metrics, organizations can establish baseline performance levels and set triggers for alert notifications when deviations occur.

bashCopy code

aws cloudwatch put-metric-alarm --alarm-name "HighCPUUtilization" --alarm-description "Alert on high CPU utilization" --metric-name "CPUUtilization" --namespace "AWS/EC2" --statistic "Average" --dimensions "Name=InstanceId,Value=i-1234567890abcdef0" --period 300 --threshold 90 --comparison-operator "GreaterThanOrEqualToThreshold" --evaluation-periods 2 --alarm-actions "arn:aws:sns:us-east-1:123456789012:MyTopic" --unit "Percent"

In this example, the AWS CLI command **aws cloudwatch put-metric-alarm** is used to create a CloudWatch alarm named "HighCPUUtilization" that monitors CPU utilization of an EC2 instance (specified by its instance ID). The alarm triggers when the average CPU utilization exceeds 90% for two consecutive evaluation periods of 5 minutes each, and it sends notifications to an SNS topic specified by its ARN.

Moreover, it's essential to consider the severity and urgency of different alerts to prioritize responses effectively. Organizations can categorize alerts into different severity levels (e.g., critical, warning, informational) based on the potential impact on their operations and define escalation procedures accordingly.

Additionally, alerting policies should incorporate contextual information to provide meaningful insights into the root causes of issues. This may include correlating metrics with application logs, infrastructure changes, deployment events, and external factors (e.g., changes in user behavior or traffic patterns). By contextualizing alerts, organizations can streamline troubleshooting and incident response processes, reducing mean time to resolution (MTTR) and minimizing business impact.

bashCopy code

```
aws cloudwatch put-metric-alarm --alarm-name "HighErrorRate" --alarm-description "Alert on high error rate in application logs" --metric-name "ErrorRate" --namespace "MyApp/Application" --statistic "Sum" --period 60 --threshold 10 --comparison-operator "GreaterThanOrEqualToThreshold" --evaluation-periods 3 --alarm-actions "arn:aws:sns:us-east-1:123456789012:MyTopic" --datapoints-to-alarm 1
```

In this example, the AWS CLI command **aws cloudwatch put-metric-alarm** is used to create a CloudWatch alarm named "HighErrorRate" that monitors the sum of error rates in application logs. The alarm triggers when the error rate exceeds 10% for three consecutive evaluation periods of 1 minute each, and it sends notifications to an SNS topic specified by its ARN.

Furthermore, organizations should continuously review and refine their alerting policies to adapt to evolving business requirements, changes in workload patterns, and advancements in technology. This iterative process involves analyzing historical alert data, gathering feedback from stakeholders, and leveraging automation and machine learning capabilities to optimize alert thresholds and reduce false positives.

In summary, designing alerting policies for AWS infrastructure requires careful consideration of key metrics, thresholds, severity levels, contextual information, and ongoing optimization efforts. By implementing effective alerting policies, organizations can proactively monitor their cloud environments, detect and respond to issues promptly, and maintain the reliability and performance of their applications and services.

Chapter 8: Managing Multi-Region Deployments with Terraform

Multi-region deployment patterns with Terraform enable organizations to distribute their infrastructure and applications across multiple AWS regions for improved availability, fault tolerance, and latency optimization. By deploying resources in multiple regions, organizations can mitigate the impact of regional outages, comply with data residency requirements, and provide better performance to users located in different geographic locations.

bashCopy code

terraform workspace new us-east-1

One of the key techniques for multi-region deployments is leveraging Terraform workspaces to manage separate configurations for each region. Workspaces allow users to maintain distinct state files for different environments, such as development, staging, and production, enabling parallel deployment and testing of infrastructure across multiple regions.

bashCopy code

terraform workspace select us-west-2

To create a multi-region deployment, organizations typically define their infrastructure configurations using Terraform modules, which encapsulate reusable components and best practices. These modules abstract the complexity of managing resources across different regions, providing a consistent and scalable approach to deploying infrastructure.

bashCopy code

terraform apply -var-file=us-east-1.tfvars

When deploying resources across multiple regions, it's essential to consider network connectivity and data replication requirements. Organizations can use AWS services like VPC peering, VPN connections, or AWS Direct Connect to establish secure communication between regions and replicate data across regional boundaries for disaster recovery and high availability purposes.

bashCopy code

terraform apply -var-file=us-west-2.tfvars

Additionally, organizations should implement strategies for managing DNS routing and traffic distribution across multiple regions. AWS Route 53, the cloud DNS service, offers features like latency-based routing and geolocation routing, allowing organizations to direct users to the nearest available region based on their geographic location or network latency.

bashCopy code

terraform apply

Moreover, organizations should design their applications with multi-region resiliency in mind, implementing techniques like active-active or active-passive architectures to ensure continuous availability and performance optimization. This may involve deploying redundant resources in multiple regions and implementing failover mechanisms to automatically redirect traffic in case of regional failures.

bashCopy code

```
terraform destroy
```

Monitoring and observability are critical aspects of managing multi-region deployments. Organizations should implement comprehensive monitoring solutions using tools like Amazon CloudWatch, AWS X-Ray, and third-party monitoring services to track the health, performance, and availability of resources across regions. By monitoring key metrics and setting up alerts, organizations can proactively detect and respond to issues in real-time, minimizing downtime and customer impact.

bashCopy code

```
terraform workspace select us-east-1
```

Furthermore, organizations should regularly test their multi-region deployments through automated testing frameworks and chaos engineering practices to validate the effectiveness of failover mechanisms and disaster recovery strategies. By simulating regional failures and conducting regular drills, organizations can identify and address potential weaknesses in their deployment architectures, ensuring readiness for real-world scenarios.

bashCopy code

```
terraform workspace delete us-east-1
```

In summary, multi-region deployment patterns with Terraform offer organizations a robust approach to building resilient and scalable infrastructure across geographically distributed AWS regions. By leveraging Terraform workspaces, modular design principles, networking capabilities, and application resilience techniques, organizations can achieve high availability, fault tolerance, and performance optimization for their cloud-based applications and services.

Implementing global infrastructure with AWS services involves designing and deploying cloud resources across multiple AWS regions to achieve high availability, fault tolerance, and low latency for users worldwide. This approach allows organizations to distribute their workload across geographically diverse data centers, ensuring resilience against regional outages and providing optimal performance to users regardless of their location.

To begin implementing global infrastructure on AWS, organizations typically start by selecting the regions where they want to deploy their resources. AWS provides a wide range of regions globally, allowing organizations to choose locations that are geographically close to their users or comply with data residency requirements.

bashCopy code

```
aws ec2 describe-regions
```

Once the regions are chosen, organizations can use AWS services like Amazon Route 53, the domain name system (DNS) service, to route traffic to the nearest available region based on users' geographic locations or network latency. Route 53 supports various routing policies, including latency-based routing, geolocation routing, and weighted routing, enabling organizations to implement sophisticated traffic distribution strategies.

bashCopy code

```
aws route53 create-hosted-zone --name example.com --caller-reference 2024-02-26-01
```

In addition to DNS routing, organizations can leverage AWS Global Accelerator to improve the availability and performance of their applications by using the AWS global network infrastructure. Global Accelerator intelligently routes user traffic to the optimal endpoint based on health, proximity, and network conditions, providing a fast and reliable experience for users worldwide.

bashCopy code

```
aws globalaccelerator create-accelerator --name MyAccelerator --ip-address-type IPV4
```

To deploy resources across multiple regions, organizations can use AWS CloudFormation or AWS CDK (Cloud Development Kit) to define their infrastructure as code and manage it in a scalable and reproducible manner. CloudFormation templates or CDK scripts allow organizations to specify the desired state of their resources and automate the deployment process across multiple regions with ease.

bashCopy code

```
aws cloudformation create-stack --stack-name MyStack --template-body file://my-template.yaml --region us-east-1
```

When deploying global infrastructure, organizations should consider data replication and synchronization requirements to ensure consistency and data durability across regions. AWS provides services like Amazon S3 (Simple Storage Service) and Amazon DynamoDB, which offer built-in capabilities for cross-region replication and data consistency, enabling organizations to replicate data seamlessly across multiple regions.

bashCopy code

```
aws s3 mb s3://my-bucket --region us-west-2
```

Furthermore, organizations should implement disaster recovery strategies to mitigate the impact of regional failures and ensure business continuity. AWS services like AWS Backup and AWS Disaster Recovery offer automated backup and recovery solutions, allowing organizations to replicate their data and applications to secondary regions and quickly recover from disasters with minimal downtime.

bashCopy code

```
aws backup create-backup-vault --backup-vault-name MyBackupVault --region us-west-1
```

Monitoring and managing global infrastructure is essential for maintaining its health, performance, and security. AWS provides services like Amazon CloudWatch, AWS Config, and AWS Trusted Advisor, which offer real-time monitoring, configuration management, and security best practices recommendations, enabling organizations to ensure compliance and optimize their global infrastructure's operation.

```bash
bashCopy code
aws cloudwatch put-metric-alarm --alarm-name MyAlarm --metric-name CPUUtilization --namespace AWS/EC2 --statistic Average --comparison-operator GreaterThanThreshold --threshold 90 --period 300 --evaluation-periods 2 --alarm-actions arn:aws:sns:us-east-1:123456789012:MyTopic
```

In summary, implementing global infrastructure with AWS services enables organizations to build highly available, fault-tolerant, and performant architectures that cater to users worldwide. By leveraging AWS's global network infrastructure, DNS routing capabilities, infrastructure as code tools, data replication mechanisms, disaster recovery solutions, and monitoring services, organizations can create resilient and scalable infrastructure that meets the demands of modern cloud applications and services.

Chapter 9: Scaling Database Workloads on AWS with Terraform

Managing database scaling with Terraform involves dynamically adjusting the capacity of databases to accommodate changes in workload demand and optimize performance. This process enables organizations to efficiently scale their databases up or down based on factors such as increased traffic, data volume, or performance requirements. Terraform provides a robust infrastructure as code (IaC) approach to automate database scaling operations, allowing organizations to define and manage their database resources programmatically.

When managing database scaling with Terraform, organizations typically start by defining the database resources they want to provision using Terraform configuration files. These configuration files, written in HashiCorp Configuration Language (HCL) or using the AWS Cloud Development Kit (CDK), specify the desired state of the database resources, including their type, size, and configuration settings.

bashCopy code

terraform { required_providers { aws = { source = "hashicorp/aws" version = "~> 3.0" } } } provider "aws" { region = "us-east-1" } resource "aws_db_instance" "example" { allocated_storage =

20 engine = "mysql" engine_version = "5.7" instance_class = "db.t2.micro" name = "mydb" username = "foo" password = "bar" parameter_group_name = "default.mysql5.7" }

Once the database resources are defined, organizations can use Terraform commands to create, update, or delete these resources based on changes to the Terraform configuration files. For example, to apply the changes defined in the Terraform configuration and provision the database resources, organizations can use the **terraform apply** command.

bashCopy code

terraform apply

To scale a database up or down using Terraform, organizations can modify the configuration settings of the database resource, such as the instance class or storage size, in the Terraform configuration files. After making the necessary changes, organizations can apply the changes using the **terraform apply** command to update the database resources accordingly.

bashCopy code

terraform apply

Terraform's state management feature keeps track of the current state of the infrastructure and enables organizations to manage changes to their infrastructure in a safe and predictable manner. When scaling database resources with Terraform, changes to the infrastructure are applied incrementally, ensuring that only the necessary modifications are made to achieve the desired state.

In addition to scaling database resources manually, organizations can implement automated scaling policies using AWS services such as Amazon RDS (Relational Database Service) or Amazon Aurora. These services offer built-in features for automated database scaling based on metrics such as CPU utilization, storage capacity, or incoming connections.

For example, organizations can configure auto-scaling policies for their Amazon RDS instances using the AWS Management Console or the AWS CLI. By defining scaling policies based on predefined thresholds or custom metrics, organizations can automatically adjust the capacity of their database instances to handle changes in workload demand.

bashCopy code

```
aws rds modify-db-instance --db-instance-identifier mydbinstance --allocated-storage 50
```

Furthermore, organizations can use Terraform to define and manage auto-scaling policies for their database resources as part of their infrastructure as code (IaC) practices. By incorporating auto-scaling configurations into their Terraform templates, organizations can ensure that their database resources scale automatically in response to changes in workload demand without manual intervention.

bashCopy code

```
resource    "aws_db_instance"    "example"    {    ...
monitoring_interval              =            60
auto_minor_version_upgrade       =            true
performance_insights_enabled = true }
```

In summary, managing database scaling with Terraform enables organizations to automate the process of adjusting database capacity to meet changing workload demands effectively. By defining database resources as code and incorporating auto-scaling configurations into their Terraform templates, organizations can ensure that their databases scale seamlessly and efficiently, optimizing performance and resource utilization in the cloud environment.

Implementing database replication and sharding strategies is crucial for ensuring high availability, scalability, and reliability of databases in modern cloud environments. Database replication involves creating and maintaining multiple copies of the same database across different servers or regions to distribute workload and provide fault tolerance. On the other hand, sharding is a technique used to horizontally partition a database into smaller, more manageable subsets called shards, each handling a portion of the overall workload. Together, these strategies enable organizations to scale their databases horizontally and vertically, improve performance, and minimize downtime.

To implement database replication using Terraform, organizations can define replication configurations in their Terraform templates to provision and manage the necessary resources automatically. For example, to set up replication for an Amazon RDS (Relational Database Service) instance, organizations can use Terraform to create a read replica of the primary database instance across multiple availability zones or regions.

bashCopy code

```
resource "aws_db_instance" "primary" { ... }
resource "aws_db_instance" "replica" { ...
replication_source_identifier =
aws_db_instance.primary.id }
```

By defining replication configurations in Terraform, organizations can easily scale their databases horizontally by adding read replicas and distributing read traffic across multiple instances. Additionally, Terraform's state management feature ensures that changes to the replication setup are applied consistently and accurately across all environments.

In addition to replication, organizations can implement sharding strategies using Terraform to partition their databases into smaller shards and distribute data more evenly across multiple instances. With Terraform, organizations can define and manage the infrastructure required for sharding, including the creation of database shards, routing tables, and load balancers.

bashCopy code

resource "aws_db_instance" "shard1" { ... } resource "aws_db_instance" "shard2" { ... } resource "aws_lb" "database_lb" { ... } resource "aws_lb_target_group" "shard1_targets" { ... } resource "aws_lb_target_group" "shard2_targets" { ... } resource "aws_lb_listener" "database_listener" { ... } By partitioning the database into smaller shards and distributing data across multiple instances, organizations can improve the scalability and performance of their databases, as each shard can handle a subset of the overall workload independently. Moreover, sharding provides fault isolation, as issues with one shard are less likely to impact the entire database system.

Furthermore, organizations can combine database replication and sharding strategies to achieve both high availability and scalability for their databases. By replicating each shard across multiple instances and regions, organizations can ensure data redundancy and fault tolerance, while sharding enables them to distribute workload effectively and scale horizontally as needed.

To deploy database replication and sharding strategies effectively, organizations should carefully design their database architecture, taking into account factors such as data consistency, latency, and access patterns. They should also monitor and optimize their database performance regularly to ensure that replication and sharding configurations meet their performance and scalability requirements. In summary, implementing database replication and sharding strategies with Terraform enables organizations to achieve high availability, scalability, and reliability for their databases in cloud environments. By automating the provisioning and management of replication and sharding configurations, Terraform streamlines the deployment process and ensures consistency across environments, allowing organizations to focus on building resilient and scalable applications.

Chapter 10: Compliance and Governance Automation with Terraform

Implementing compliance policies with Terraform is crucial for organizations aiming to maintain regulatory compliance and security standards across their cloud infrastructure. Compliance policies define a set of rules and guidelines that organizations must adhere to ensure their infrastructure meets specific regulatory requirements, such as HIPAA, GDPR, PCI DSS, or SOC 2. By leveraging Terraform's infrastructure as code (IaC) capabilities, organizations can automate the enforcement of compliance policies, ensuring consistent and auditable infrastructure configurations.

To implement compliance policies with Terraform, organizations first need to define the compliance requirements relevant to their industry and jurisdiction. This involves understanding the regulatory standards applicable to their organization and identifying the specific controls and configurations necessary to achieve compliance. Once the compliance requirements are defined, organizations can translate them into Terraform configurations to automate the provisioning and management of compliant infrastructure.

For example, suppose an organization needs to comply with the Payment Card Industry Data Security Standard (PCI DSS), which requires restricting access to cardholder data and encrypting sensitive information. In that case, they can use Terraform to enforce compliance by defining security groups, network ACLs, and encryption settings for their cloud resources.

bashCopy code

```
resource "aws_security_group" "pci_security_group" {
... ingress { from_port = 443 to_port = 443 protocol = "tcp" cidr_blocks = ["0.0.0.0/0"] } egress { from_port = 0 to_port = 0 protocol = "-1" cidr_blocks = ["0.0.0.0/0"] } } resource "aws_db_instance" "pci_database" { ... kms_key_id = aws_kms_key.pci_key.arn }
```

In this example, the Terraform configuration defines a security group that allows inbound traffic on port 443 (HTTPS) and restricts outbound traffic to prevent unauthorized access. Additionally, the configuration specifies an AWS RDS instance encrypted using a customer-managed KMS (Key Management Service) key to protect sensitive data at rest.

By defining compliance policies as code with Terraform, organizations can automate the enforcement of security controls and configurations, reducing the risk of human error and ensuring consistent compliance across their infrastructure. Furthermore, Terraform's declarative syntax enables organizations to define infrastructure configurations in a clear and concise manner, making it easier to understand and maintain compliance requirements over time.

To ensure ongoing compliance, organizations should regularly review and update their Terraform configurations to reflect changes in regulatory requirements and security best practices. They should also implement continuous compliance monitoring and auditing processes to detect and remediate non-compliant configurations proactively.

In summary, implementing compliance policies with Terraform enables organizations to automate the enforcement of regulatory requirements and security standards across their cloud infrastructure. By defining compliance requirements as code, organizations can ensure consistent and auditable infrastructure configurations while reducing the burden of manual configuration management. Ultimately, Terraform empowers organizations to achieve and maintain compliance efficiently and effectively in dynamic cloud environments.

Automating governance processes on AWS is essential for organizations seeking to maintain control, compliance, and security across their cloud environments. Governance encompasses a wide range of activities, including defining policies, implementing controls, auditing configurations, and enforcing compliance standards. By leveraging AWS services and automation tools, organizations can streamline governance processes, reduce manual effort, and ensure consistency and adherence to internal policies and external regulations.

One fundamental aspect of automating governance on AWS is the use of AWS Config. AWS Config enables organizations to assess, audit, and evaluate the configurations of their AWS resources continuously. By capturing resource configurations and changes over time, AWS Config provides a historical record of resource configuration states, allowing organizations to track compliance with internal policies and industry regulations.

To automate governance using AWS Config, organizations can define configuration rules that evaluate resource configurations against predefined criteria. For example, organizations can create rules to enforce encryption settings, enforce security group rules, or ensure proper tagging of resources. These rules can be defined using AWS Config's managed rules or custom rules written in AWS Lambda functions.

```bash
bashCopy code
# Enable AWS Config aws configservice put-config-rule --config-rule-name encryption-check --region us-east-1 --scope ComplianceResourceTypes='AWS::S3::Bucket' --source Owner=AWS,SourceIdentifier=s3-bucket-encryption-check --input-parameters "SNSTopicArn=<SNS_TOPIC_ARN>" # Create Custom Config Rule aws configservice put-config-rule --config-rule file://s3-bucket-encryption-rule.json
```

In this example, the first command enables AWS Config in the specified region and scopes it to assess compliance for S3 buckets. The command also specifies an SNS topic to receive notifications for non-compliant resources. The second command creates a custom configuration rule by specifying a JSON file containing the rule's configuration, such as the resource type to evaluate and the criteria for compliance.

Another essential aspect of automating governance on AWS is the use of AWS Service Catalog. AWS Service Catalog allows organizations to create and manage catalogs of approved IT services and resources. By defining product portfolios and constraints, organizations can ensure that users deploy resources that meet predefined governance requirements.

To automate governance using AWS Service Catalog, organizations can create product portfolios containing approved resources and templates. They can also define constraints to enforce compliance policies, such as restricting instance types, enforcing tagging policies, or requiring encryption settings.

bashCopy code

```
# Create Product aws servicecatalog create-product --name "MyEC2Instance" --owner "my-company" --description "EC2 instance template" --provider-name "AWS" # Create Portfolio aws servicecatalog create-portfolio --display-name "My Portfolio" --provider-name "My Company" # Associate Product with Portfolio aws servicecatalog associate-product-with-portfolio --product-id "product-id" --portfolio-id "portfolio-id"
```

In this example, the commands create a product representing an EC2 instance template and a portfolio to contain approved products. The product is associated with the portfolio, making it available for users to deploy within the organization's governance framework.

Furthermore, organizations can leverage AWS Organizations to automate governance across multiple AWS accounts. AWS Organizations enables central management of policies, service control policies (SCPs), and organizational units (OUs) to enforce compliance and security standards across the organization's AWS accounts.

bashCopy code

```
# Create Organizational Unit aws organizations create-organizational-unit --parent-id "ou-parent-id" --name "DevOU" # Attach SCP to OU aws organizations attach-policy --policy-id "policy-id" --target-id "ou-id"
```

In this example, the commands create an organizational unit for development resources and attach a service control policy (SCP) to enforce compliance requirements, such as restricting access to specific AWS services or regions.

Overall, automating governance processes on AWS using services like AWS Config, AWS Service Catalog, and AWS Organizations enables organizations to establish and enforce compliance standards efficiently. By defining policies, implementing controls, and automating enforcement, organizations can ensure the security, compliance, and reliability of their cloud environments while reducing manual effort and operational risk.

BOOK 4
EXPERT AWS CLOUD AUTOMATION
SCALING AND MANAGING COMPLEX DEPLOYMENTS
WITH TERRAFORM

ROB BOTWRIGHT

Chapter 1: Advanced Terraform Configuration Management

Dynamic configuration using Terraform variables is a fundamental aspect of infrastructure as code (IaC) that enables users to create reusable and flexible infrastructure definitions. Terraform variables allow for the parameterization of configuration values, making it possible to create dynamic and customizable infrastructure deployments. By leveraging variables, users can define their infrastructure configurations in a modular and scalable manner, enhancing maintainability and reducing redundancy.

To start using Terraform variables, users can define variables within their Terraform configuration files using the **variable** block. For example, to define a variable for the AWS region, users can create a **variables.tf** file with the following content:

hclCopy code

variable "region" { description = "The AWS region where resources will be provisioned" default = "us-west-2" }

In this example, the **variable** block defines a variable named **region** with a default value of **"us-west-2"**. Users can override this default value by specifying a value in a separate configuration file or by passing the value as a command-line argument.

bashCopy code

terraform plan -var="region=us-east-1"

This command instructs Terraform to use **"us-east-1"** as the value for the **region** variable during the planning phase. Additionally, users can specify variable values using a **terraform.tfvars** file, which Terraform automatically loads when executing commands.

hclCopy code

terraform.tfvars region = "eu-west-1"

bashCopy code

terraform plan

Terraform also supports input variables, which prompt users to provide values interactively during execution. Input variables are particularly useful when running Terraform in automation workflows or when deploying infrastructure in environments with varying configurations.

hclCopy code

variable "environment" { description = "The target environment for the infrastructure" type = string } resource "aws_instance" "example" { ... tags = { Environment = var.environment } }

In this example, the **environment** variable is defined as an input variable, prompting users to provide a value when executing Terraform commands. The value of the **environment** variable is then used to tag AWS resources created by the **aws_instance** resource.

Terraform variables support various data types, including strings, numbers, lists, maps, and boolean values, providing users with flexibility in defining their infrastructure configurations. Additionally, users can define variables at different levels of granularity, including module-level variables, allowing for the modularization and reuse of infrastructure components.

Furthermore, Terraform variables support interpolation, enabling users to reference the values of other variables or resources dynamically within their configurations. Interpolation allows for the creation of more dynamic and context-aware infrastructure definitions, enhancing the flexibility and expressiveness of Terraform configurations.

hclCopy code

```
variable "subnet_ids" { description = "A list of subnet IDs for the VPC" type = list(string) } resource "aws_instance" "example" { ... subnet_id = var.subnet_ids[count.index] }
```

In this example, the **subnet_ids** variable is defined as a list of subnet IDs, which are dynamically assigned to AWS instances created by the **aws_instance** resource using the **count.index** interpolation syntax.

Overall, Terraform variables play a crucial role in enabling dynamic and configurable infrastructure deployments. By leveraging variables, users can create modular, reusable, and flexible infrastructure definitions, empowering them to manage complex infrastructure environments with ease. Whether defining input variables for interactive deployments or using default values and interpolation for automation workflows, Terraform variables provide users with the tools they need to achieve dynamic configuration management on AWS and other cloud platforms.

Terraform's template rendering techniques provide a robust means to dynamically configure infrastructure resources based on predefined templates. These techniques are pivotal in managing complex cloud environments efficiently. One of the primary strengths of Terraform lies in its ability to abstract infrastructure as code (IaC), enabling users to define and provision resources programmatically across various cloud platforms, including AWS, Azure, and Google Cloud Platform.

At the heart of Terraform's template rendering capabilities is its interpolation syntax, which allows for dynamic referencing of variables, data sources, and outputs within configuration files. This enables users to parameterize their infrastructure definitions and adapt them to different environments or requirements seamlessly. For instance, consider a scenario where you need to define multiple similar resources with only minor variations. Instead of duplicating configuration blocks, you can leverage Terraform's template rendering to generate these resources programmatically.

The **${}** syntax is commonly used for interpolation in Terraform. This syntax allows users to embed variables directly into resource configurations. For example, to reference the value of a variable named **subnet_id**, you would use **${var.subnet_id}** within your Terraform configuration files.

Another essential aspect of Terraform's template rendering techniques is the use of functions. Terraform provides a rich set of built-in functions that enable various operations, such as string manipulation, mathematical calculations, and data formatting. These functions enhance the flexibility and expressiveness of Terraform configurations. For instance, the **format** function allows users to construct strings dynamically based on input values.

hclCopy code

```
resource "aws_instance" "example" { ami = "ami-12345678" instance_type = "t2.micro" tags = { Name = format("web-%s", var.environment) } }
```

In this example, the **format** function is used to generate dynamic tag values based on the **environment** variable.

Furthermore, Terraform supports the use of conditionals within configuration files, allowing users to define logic based on certain conditions. This is particularly useful for handling scenarios where resource configurations need to vary based on environment types or other factors.

hclCopy code

```
resource "aws_instance" "example" { ami =
var.environment == "production" ? "ami-prod" :
"ami-dev" instance_type = "t2.micro" }
```

In this snippet, the **ami** attribute of the **aws_instance** resource is conditionally set based on the value of the **environment** variable.

Additionally, Terraform's template rendering capabilities extend to the use of external data sources and outputs. Data sources allow Terraform configurations to retrieve information from external systems, such as AWS S3 buckets or DynamoDB tables, and incorporate that data into resource definitions. Outputs, on the other hand, enable users to expose certain values generated during the Terraform execution, making them accessible to other Terraform configurations or external systems.

hclCopy code

```
data "aws_vpc" "example" { id = var.vpc_id } output
"vpc_cidr_block" { value =
data.aws_vpc.example.cidr_block }
```

In this example, a data source is used to retrieve information about an existing AWS VPC, and an output is defined to expose the VPC's CIDR block.

By leveraging Terraform's template rendering techniques, users can create highly adaptable and reusable infrastructure configurations, streamlining the management of cloud resources and enhancing overall operational efficiency.

Chapter 2: Orchestrating Complex Deployments with Terraform

Implementing Terraform workflows for complex deployments is crucial for managing intricate infrastructure configurations efficiently. Terraform offers various features and best practices to streamline the deployment process and maintain infrastructure consistency across different environments. These workflows encompass multiple stages, including planning, provisioning, and managing infrastructure resources, and involve the use of modules, workspaces, and version control systems to ensure scalability, reliability, and maintainability.

One fundamental aspect of implementing Terraform workflows is organizing infrastructure code into modular components. Modularization allows for better code organization, reusability, and easier maintenance. Terraform modules encapsulate related resources into reusable units, enabling users to define infrastructure components once and reuse them across multiple projects or environments. To create a Terraform module, you can use the **terraform module create** command to generate a new module structure.

bashCopy code

terraform module create <MODULE_NAME>

Once the module structure is created, you can define the necessary resources and variables within the module directory. Modules can then be reused by referencing them in other Terraform configurations using the **module** block.

hclCopy code

```
module "example_module" { source = "./modules/example" var1 = "value1" var2 = "value2" }
```

Furthermore, Terraform workspaces facilitate managing multiple environments, such as development, staging, and production, within a single configuration. Workspaces allow users to maintain separate state files for each environment, enabling changes to be isolated and applied independently. To create a new workspace, you can use the **terraform workspace new** command.

bashCopy code

```
terraform workspace new <WORKSPACE_NAME>
```

Once the workspace is created, you can switch between workspaces using the **terraform workspace select** command.

bashCopy code

```
terraform workspace select <WORKSPACE_NAME>
```

Another critical aspect of implementing Terraform workflows is version control. Version control systems such as Git enable collaborative development, change tracking, and rollback capabilities for infrastructure code. By storing Terraform configurations in a version control repository, users can manage changes more effectively and ensure consistency across environments. To initialize a new Git repository for Terraform configurations, you can use the following commands.

bashCopy code

git init git add . git commit -m "Initial commit"

Subsequently, users can push changes to a remote repository and collaborate with team members using Git branching and merging strategies.

In addition to version control, Terraform's state management capabilities play a crucial role in implementing workflows for complex deployments. Terraform state files store information about the current state of deployed resources, enabling Terraform to track changes and manage resource dependencies. By default, Terraform stores state files locally, but for collaborative or production environments, it's recommended to use remote backends such as Amazon S3 or Terraform Cloud for centralized state management.

To configure a remote backend, you can specify the backend configuration in the Terraform configuration file.

hclCopy code

terraform { backend "s3" { bucket = "example-bucket" key = "terraform.tfstate" region = "us-east-1" } }

Furthermore, implementing automated workflows with Terraform involves integrating with continuous integration/continuous deployment (CI/CD) pipelines. CI/CD pipelines automate the process of building, testing, and deploying infrastructure changes, reducing manual intervention and ensuring consistent deployments. Popular CI/CD platforms such as Jenkins, GitLab CI/CD, and GitHub Actions support Terraform integration, allowing users to trigger Terraform commands within pipeline workflows.

For instance, in a Jenkins pipeline, you can use the **terraform plan** and **terraform apply** commands to perform infrastructure changes.

groovyCopy code

```
pipeline { agent any stages { stage('Terraform Plan') { steps { sh 'terraform plan -out=tfplan' } } stage('Terraform Apply') { steps { sh 'terraform apply tfplan' } } } }
```

By implementing Terraform workflows for complex deployments, organizations can achieve greater efficiency, reliability, and scalability in managing their infrastructure as code. These workflows incorporate best practices such as modularization, workspace management, version control, state management, and CI/CD integration to streamline the deployment process and maintain infrastructure consistency across various environments.

Managing dependencies and interactions in

Terraform is essential for ensuring the smooth deployment and operation of infrastructure. Terraform allows users to define dependencies between resources and manage their interactions through various techniques such as resource dependencies, provisioners, and lifecycle hooks. By understanding and effectively managing dependencies, users can ensure that resources are provisioned in the correct order, avoid race conditions, and maintain the integrity of their infrastructure.

One of the fundamental aspects of managing dependencies in Terraform is defining resource dependencies explicitly within the configuration. Terraform uses the concept of implicit and explicit dependencies to determine the order in which resources are provisioned. Implicit dependencies are automatically inferred based on resource references in the configuration, while explicit dependencies are explicitly defined using the **depends_on** attribute.

For example, consider a scenario where an EC2 instance depends on a VPC and a security group. You can define the dependencies as follows:

hclCopy code

resource "aws_vpc" "example_vpc" { // VPC configuration } resource "aws_security_group" "example_security_group" { // Security group configuration } resource "aws_instance" "example_instance" { // Instance configuration //

Explicit dependencies depends_on = [aws_vpc.example_vpc, aws_security_group.example_security_group,] }

By specifying explicit dependencies, Terraform ensures that the VPC and security group are created before provisioning the EC2 instance.

Another technique for managing dependencies and interactions is using provisioners to execute scripts or commands on provisioned resources. Provisioners allow users to perform tasks such as installing software, configuring applications, or running tests on provisioned instances. Terraform supports various provisioners, including local-exec, remote-exec, and file provisioners.

For example, to execute a local script on a provisioned EC2 instance, you can use the **remote-exec** provisioner as follows:

hclCopy code

```
resource "aws_instance" "example_instance" { // Instance configuration provisioner "remote-exec" { inline = [ "sudo apt-get update", "sudo apt-get install -y nginx", ] } }
```

This provisioner executes the specified commands on the provisioned EC2 instance after it is created.

Furthermore, Terraform provides lifecycle hooks to control the behavior of resources during certain lifecycle events, such as creation, update, and deletion. Lifecycle hooks allow users to customize resource behavior by specifying actions to be performed before or after certain lifecycle events.

For instance, you can use the **create_before_destroy** lifecycle configuration to create a new resource before destroying the existing one, ensuring zero downtime during updates.

hclCopy code

```
resource "aws_instance" "example_instance" { // Instance configuration lifecycle { create_before_destroy = true } }
```

This configuration instructs Terraform to create a new instance before destroying the existing one during updates.

In addition to these techniques, Terraform modules provide a powerful mechanism for managing dependencies and interactions between infrastructure components. Modules encapsulate related resources and configurations into reusable units, allowing users to define complex infrastructure patterns and manage dependencies at a higher level of abstraction.

For example, you can create a module for provisioning a web server stack, including instances, security groups, and load balancers, and then reuse this module across multiple projects or environments.

hclCopy code

```
module "web_server_stack" { source = "./modules/web_server_stack" // Module inputs }
```

By encapsulating dependencies within modules, users can simplify configuration management, promote reusability, and ensure consistency across deployments.

In summary, managing dependencies and interactions in Terraform is crucial for orchestrating complex infrastructure deployments effectively. By defining explicit dependencies, using provisioners, leveraging lifecycle hooks, and utilizing modules, users can ensure that resources are provisioned in the correct order, execute necessary tasks on provisioned resources, customize resource behavior, and promote reusability and consistency across deployments.

Chapter 3: Managing Microservices Architecture on AWS with Terraform

Designing microservices architecture with Terraform involves orchestrating the deployment and management of a distributed system composed of loosely coupled and independently deployable services. Microservices architecture aims to break down monolithic applications into smaller, more manageable services that can be developed, deployed, and scaled independently. Terraform provides a powerful toolset for designing and provisioning the infrastructure required to support microservices architecture efficiently.

The first step in designing microservices architecture with Terraform is to identify the individual services that make up the application and define their boundaries. Each microservice should have a clear purpose and responsibility, and communication between services should be through well-defined APIs. Terraform enables users to define infrastructure as code, allowing them to declare the resources needed to support each microservice.

To begin designing microservices architecture with Terraform, users typically create a separate Terraform configuration for each microservice. Within each configuration, users define the resources required for the microservice, such as compute instances, containers, databases, networking components, and security policies. For example, to provision a microservice that serves web requests, users may define an AWS ECS (Elastic Container Service) task definition, an application load balancer, and an auto-scaling group using Terraform:

hclCopy code

```
resource "aws_ecs_task_definition" "web_service" {
// Task definition configuration } resource
"aws_lb_target_group" "web_service_target_group"
{ // Target group configuration } resource
"aws_lb_listener_rule" "web_service_rule" { //
Listener rule configuration } resource
"aws_autoscaling_group"
"web_service_autoscaling_group" { // Auto-scaling
group configuration }
```

Once the Terraform configurations for each microservice are defined, users can use Terraform CLI commands to initialize, plan, and apply the infrastructure changes. For example, to initialize the Terraform workspace and download provider plugins, users can run:

bashCopy code

```
terraform init
```

To preview the changes Terraform will make to the infrastructure, users can run:

bashCopy code

```
terraform plan
```

And to apply the changes and provision the infrastructure, users can run:

bashCopy code

terraform apply

By organizing microservices into separate Terraform configurations, users can manage each service's infrastructure independently, enabling teams to iterate, scale, and deploy services more efficiently.

In addition to defining the infrastructure for individual microservices, Terraform allows users to manage cross-cutting concerns such as networking, security, monitoring, and logging. For example, users can define a Terraform module for creating a shared VPC (Virtual Private Cloud) and security groups to ensure consistent network isolation and security policies across microservices.

hclCopy code

module "shared_vpc" { source = "./modules/shared_vpc" // Module inputs }

By modularizing infrastructure components, users can promote reusability, consistency, and maintainability across microservices.

Furthermore, Terraform integrates seamlessly with other tools and services commonly used in microservices architecture, such as container orchestration platforms like Kubernetes or AWS ECS, service discovery tools like Consul or AWS Route 53, and continuous integration and delivery (CI/CD) pipelines like Jenkins or AWS CodePipeline. Users can leverage Terraform's provider ecosystem to interact with these services and automate the provisioning and management of microservices infrastructure.

Another important aspect of designing microservices architecture with Terraform is managing service dependencies and interactions. Microservices often rely on other services or external dependencies, and orchestrating these interactions is critical for ensuring the reliability and resilience of the system. Terraform allows users to define dependencies between resources explicitly using the **depends_on** attribute and manage service interactions through provisioners, lifecycle hooks, and other advanced features.

In summary, designing microservices architecture with Terraform involves defining the infrastructure required for each microservice, managing cross-cutting concerns, integrating with other tools and services, and orchestrating service dependencies and interactions. By leveraging Terraform's infrastructure as code capabilities, users can design scalable, resilient, and maintainable microservices architectures that enable rapid iteration and deployment of applications.

Deploying and scaling microservices with Terraform and AWS involves orchestrating the infrastructure required to support a distributed system of independently deployable services and dynamically scaling them based on demand. Terraform provides a powerful infrastructure as code toolset, while AWS offers a comprehensive set of services for building scalable and resilient microservices architectures.

The first step in deploying and scaling microservices with Terraform and AWS is to define the infrastructure for each microservice using Terraform configurations. Users typically create separate Terraform configurations for each microservice, specifying the resources needed to support the service, such as compute instances, containers, databases, networking components, and security policies. For example, to deploy a microservice using AWS ECS (Elastic Container Service), users can define a Terraform configuration that includes an ECS task definition, a service definition, and associated networking components:

hclCopy code

```
resource "aws_ecs_task_definition" "my_microservice_task" { // Task definition configuration } resource "aws_ecs_service" "my_microservice_service" { // Service definition configuration } resource "aws_lb_target_group" "my_microservice_target_group" { // Target group configuration } resource "aws_lb_listener_rule" "my_microservice_rule" { // Listener rule configuration }
```

Once the Terraform configurations for each microservice are defined, users can use Terraform CLI commands to initialize, plan, and apply the infrastructure changes. For example, to initialize the Terraform workspace and download provider plugins, users can run:

bashCopy code

terraform init

To preview the changes Terraform will make to the infrastructure, users can run:

bashCopy code

terraform plan

And to apply the changes and provision the infrastructure, users can run:

bashCopy code

terraform apply

By following this approach, users can deploy the infrastructure required to support their microservices architecture in a repeatable and consistent manner. Terraform enables infrastructure changes to be versioned, tested, and audited, providing confidence in the deployment process.

Once the microservices are deployed, the next step is to configure auto-scaling to dynamically adjust the number of service instances based on demand. AWS offers several auto-scaling solutions, including AWS Auto Scaling, Amazon ECS Auto Scaling, and Amazon EC2 Auto Scaling, which can be integrated with Terraform to automate the scaling process.

For example, to configure auto-scaling for an ECS service using Terraform, users can define an autoscaling policy and attach it to the ECS service:

```hcl
hclCopy code
resource "aws_appautoscaling_target" "my_microservice_target" { // Target configuration }
resource "aws_appautoscaling_policy" "my_microservice_policy" { // Policy configuration }
resource "aws_ecs_service" "my_microservice_service" { // Service definition configuration depends_on = [aws_appautoscaling_target.my_microservice_target ] }
```

By specifying appropriate scaling policies based on metrics such as CPU utilization or request count, users can ensure that their microservices infrastructure automatically scales up or down to handle changes in workload, optimizing resource utilization and cost efficiency.

In addition to auto-scaling, users can leverage AWS features such as Elastic Load Balancing (ELB) and Amazon Route 53 for load balancing and traffic management, ensuring that incoming requests are distributed evenly across service instances and enabling high availability and fault tolerance.

Furthermore, Terraform provides support for managing service dependencies and interactions, enabling users to define relationships between microservices and configure communication between them. This includes setting up VPC peering connections, configuring security groups and network

ACLs, and managing DNS records to enable service discovery and communication.

Overall, deploying and scaling microservices with Terraform and AWS involves defining the infrastructure for each service using Terraform configurations, configuring auto-scaling to dynamically adjust capacity based on demand, leveraging AWS features for load balancing and traffic management, and managing service dependencies and interactions. By following best practices for infrastructure as code and leveraging the capabilities of Terraform and AWS, users can build scalable, resilient, and cost-effective microservices architectures that meet the needs of their applications.

Chapter 4: Infrastructure as Code Governance at Scale

Implementing governance policies with Terraform involves defining and enforcing rules and standards for managing infrastructure resources across an organization's cloud environment. Terraform provides capabilities to codify governance policies, ensuring consistency, compliance, and security in infrastructure deployments.

To start implementing governance policies with Terraform, organizations typically establish a set of rules and standards that align with their compliance requirements, security policies, and operational objectives. These policies may include guidelines for resource naming conventions, tagging strategies, access control configurations, encryption requirements, and compliance checks.

Once the governance policies are defined, organizations use Terraform configurations to codify these policies as code. This involves creating Terraform modules or configurations that enforce the desired policies when provisioning infrastructure resources. For example, to enforce tagging policies for resources provisioned in AWS, organizations can create Terraform configurations that include tag enforcement rules using the **aws_instance** resource:

hclCopy code

```
resource "aws_instance" "example" { // Instance
configuration tags = { Name = "ExampleInstance"
Environment = var.environment Owner = var.owner //
Add more required tags } }
```

In this example, the **tags** block specifies the required tags for the AWS instance, such as **Name, Environment,** and **Owner.** By including these tags in the Terraform configuration, organizations ensure that all instances provisioned through Terraform adhere to the tagging policy.

Additionally, organizations can use Terraform to enforce access control policies by managing IAM (Identity and Access Management) resources. For example, organizations can create Terraform configurations that define IAM policies and roles with granular permissions, ensuring that only authorized users and applications have access to resources. Here's an example of defining an IAM policy with Terraform:

hclCopy code

```
resource "aws_iam_policy" "example_policy" { name =
"ExamplePolicy" description = "Example IAM Policy"
policy = data.aws_iam_policy_document.example.json }
data "aws_iam_policy_document" "example" {
statement { effect = "Allow" actions = ["s3:GetObject"]
resources = ["arn:aws:s3:::example-bucket/*"] } }
```

This Terraform configuration creates an IAM policy that allows read access to objects in a specific S3 bucket. By managing IAM policies and roles with Terraform, organizations can ensure that access to resources is restricted according to their security policies and compliance requirements.

Furthermore, organizations can leverage Terraform to implement compliance checks and audit trails by integrating with external tools and services. For example, Terraform can be integrated with AWS Config to evaluate compliance with predefined rules and automatically remediate non-compliant resources. Organizations can use Terraform configurations to define AWS Config rules and remediation actions, ensuring continuous compliance with regulatory requirements.

To enforce governance policies in Terraform, organizations can use Terraform Enterprise features such as Sentinel, which allows organizations to define and enforce policies as code. Sentinel policies can be applied to Terraform configurations to prevent non-compliant changes from being applied to the infrastructure. Organizations can define Sentinel policies that validate resource configurations, enforce tagging standards, and restrict access to sensitive resources, providing an additional layer of governance and control.

In summary, implementing governance policies with Terraform involves defining rules and standards for managing infrastructure resources, codifying these policies as Terraform configurations or modules, enforcing access control policies with IAM resources, integrating with external tools for compliance checks and audit trails, and leveraging Terraform Enterprise features such as Sentinel for policy enforcement. By following these best practices, organizations can ensure consistency, compliance, and security in their infrastructure deployments using Terraform.

Automated compliance checks and enforcement are crucial aspects of maintaining a secure and compliant cloud infrastructure environment. Leveraging automation tools like Terraform, organizations can streamline the process of ensuring adherence to regulatory requirements, security policies, and industry standards.

To implement automated compliance checks and enforcement with Terraform, organizations typically follow a structured approach. First, they define compliance standards and requirements based on applicable regulations such as GDPR, HIPAA, PCI DSS, or internal security policies. These standards may include guidelines for resource configuration, access control, data encryption, network security, and more.

Once the compliance standards are established, organizations use Terraform to codify these requirements as infrastructure-as-code (IaC) configurations. Terraform enables organizations to define resource configurations, access policies, and security settings in a declarative manner, ensuring consistent and repeatable deployments.

For example, organizations can use Terraform to enforce compliance with encryption standards by configuring AWS resources to use encryption by default. This can be achieved using Terraform configurations that specify encryption settings for AWS services such as S3 buckets, EBS volumes, and RDS databases:

```
hclCopy code
resource "aws_s3_bucket" "example_bucket" { bucket = "example-bucket" acl = "private" server_side_encryption_configuration { rule { apply_server_side_encryption_by_default { sse_algorithm = "AES256" } } } }
```

In this example, the Terraform configuration ensures that the created S3 bucket is encrypted using AES256 encryption by default, thereby meeting encryption compliance requirements.

Additionally, organizations can use Terraform to enforce access control policies by managing IAM roles, policies, and permissions. By defining IAM policies as code, organizations can ensure that only authorized users and services have access to resources, reducing the risk of unauthorized access and data breaches.

Terraform also facilitates the implementation of compliance checks through integration with AWS Config. AWS Config enables organizations to assess compliance with predefined rules and configurations and automatically remediate non-compliant resources using Terraform. Organizations can define AWS Config rules as Terraform configurations and specify remediation actions to be taken when violations are detected.

```
hclCopy code
resource "aws_config_config_rule" "example_rule" { name = "example-rule" source { owner = "AWS"
```

```
source_identifier = "S3_BUCKET_ENCRYPTION_CHECK"
} # Specify remediation action here }
```

In this example, the Terraform configuration defines an AWS Config rule that checks whether S3 buckets are encrypted. If a non-compliant bucket is detected, Terraform can be configured to automatically remediate the issue by applying encryption settings.

Furthermore, organizations can use Terraform Enterprise features such as Sentinel for policy enforcement. Sentinel enables organizations to define and enforce compliance policies as code, ensuring that infrastructure deployments comply with regulatory requirements and security standards. Sentinel policies can be applied to Terraform configurations to prevent non-compliant changes from being applied to the infrastructure.

By leveraging automated compliance checks and enforcement with Terraform, organizations can ensure that their cloud infrastructure adheres to regulatory requirements, security policies, and industry standards. This approach helps organizations reduce manual effort, mitigate compliance risks, and maintain a secure and compliant environment at scale.

Chapter 5: Implementing Infrastructure Pipelines with Terraform

Building CI/CD pipelines for infrastructure as code (IaC) is essential for automating the deployment, testing, and delivery of infrastructure changes. By implementing CI/CD pipelines, organizations can achieve faster and more reliable infrastructure deployments, improved collaboration among development and operations teams, and better overall management of infrastructure changes.

The first step in building CI/CD pipelines for IaC is to select a suitable CI/CD tool. Popular choices include Jenkins, GitLab CI/CD, CircleCI, Travis CI, and AWS CodePipeline. Each of these tools offers features and integrations that support the automation of IaC workflows.

For instance, if using Jenkins, organizations can set up Jenkins jobs to trigger Terraform commands based on changes to the version control system. This can be achieved by configuring a Jenkinsfile, which defines the pipeline stages and tasks, including fetching the Terraform code, initializing Terraform, applying changes, and running tests.

groovyCopy code

pipeline { agent any stages { stage('Fetch Terraform code') { steps { git branch: 'main', url: 'https://github.com/your-repo/terraform.git' } }

```
stage('Initialize Terraform') { steps { sh 'terraform init' }
} stage('Apply Terraform changes') { steps { sh
'terraform apply -auto-approve' } } stage('Run tests') {
steps { // Run Terraform validation tests sh 'terraform
validate' // Run other tests as needed } } } }
```

In this Jenkinsfile example, each stage represents a phase in the CI/CD pipeline, such as fetching the Terraform code, initializing Terraform, applying changes, and running tests. The 'sh' command is used to execute Terraform CLI commands within the Jenkins pipeline.

Similarly, in GitLab CI/CD, organizations can define CI/CD pipelines using a .gitlab-ci.yml file. This file specifies the stages and jobs to be executed, including tasks for fetching the Terraform code, initializing Terraform, applying changes, and running tests.

yamlCopy code

```
stages: - build - test - deploy terraform: stage:
deploy script: - terraform init - terraform plan -
out=tfplan - terraform apply -auto-approve tfplan
```

In this GitLab CI/CD example, the 'terraform' job is defined to execute Terraform commands within the deploy stage. The 'terraform init', 'terraform plan', and 'terraform apply' commands are used to initialize Terraform, generate an execution plan, and apply changes, respectively.

Additionally, organizations can integrate their CI/CD pipelines with version control systems such as Git or AWS CodeCommit to automatically trigger pipeline executions upon code commits or pull requests. This ensures that infrastructure changes are continuously tested and deployed in a controlled manner.

Furthermore, organizations can implement automated testing and validation checks within their CI/CD pipelines to ensure the quality and reliability of infrastructure changes. This may include running Terraform validation tests, performing infrastructure linting, and executing integration tests against the deployed infrastructure.

Overall, building CI/CD pipelines for infrastructure as code enables organizations to automate the deployment, testing, and delivery of infrastructure changes, leading to increased efficiency, reduced manual effort, and improved infrastructure reliability. By leveraging CI/CD tools and best practices, organizations can streamline their IaC workflows and accelerate their journey towards infrastructure automation.

Automated testing and validation in infrastructure pipelines play a crucial role in ensuring the reliability, security, and efficiency of infrastructure deployments. By implementing automated testing and validation processes, organizations can detect and prevent potential issues early in the development lifecycle, leading to fewer errors and faster feedback loops.

One common approach to automated testing and validation in infrastructure pipelines is to integrate various testing tools and frameworks into the CI/CD workflow. For example, tools like Terratest, Kitchen-Terraform, and InSpec provide capabilities for testing

Terraform configurations, infrastructure deployments, and compliance requirements.

Terratest, a popular Go library for testing Terraform code, allows developers to write automated tests to validate infrastructure changes. Tests can be written using Go and executed as part of the CI/CD pipeline to verify the correctness of Terraform configurations and ensure that infrastructure resources are provisioned as expected.

```go
goCopy code
package test import ( "testing" "github.com/gruntwork-io/terratest/modules/terraform" "github.com/stretchr/testify/assert" ) func TestTerraformDeployment(t *testing.T) { terraformOptions := &terraform.Options{ TerraformDir: "../terraform-code", } defer terraform.Destroy(t, terraformOptions) terraform.InitAndApply(t, terraformOptions) instanceIP := terraform.Output(t, terraformOptions, "instance_ip") expectedIP := "10.0.0.1" assert.Equal(t, expectedIP, instanceIP) }
```

In this example, a Terraform deployment test is written using Terratest. The test initializes Terraform, applies changes, retrieves the output values, and asserts that the output matches the expected result. This test can be executed as part of the CI/CD pipeline to validate infrastructure changes automatically.

Similarly, Kitchen-Terraform is a testing framework that integrates with Terraform and allows developers to write tests in Ruby. Kitchen-Terraform provisions infrastructure resources using Terraform and then runs InSpec tests to verify the desired state of the infrastructure.

rubyCopy code

.kitchen.yml --- driver: name: terraform provisioner: name: terraform verifier: name: terraform systems: - name: default backend: local patterns: - test/integration/default/*_test.rb

In this configuration file, Kitchen-Terraform is configured to use InSpec tests located in the 'test/integration/default' directory to verify the deployed infrastructure. These tests can include checks for security compliance, configuration correctness, and resource availability.

Additionally, InSpec, a powerful compliance testing framework, can be used to define and enforce compliance policies for infrastructure deployments. InSpec tests are written in a human-readable format and can be executed against deployed infrastructure to validate compliance with security standards, regulatory requirements, and organizational policies.

rubyCopy code

example InSpec test describe aws_security_group('my-security-group') do it { should allow_in(port: 22, ipv4_range: '0.0.0.0/0') } end

In this example, an InSpec test verifies that a security group named 'my-security-group' allows inbound traffic on port 22 from any IPv4 address. By writing and executing such tests as part of the CI/CD pipeline, organizations can ensure that infrastructure deployments comply with security policies and regulations.

Furthermore, organizations can leverage infrastructure linting tools like Checkov and TFLint to perform static analysis of Terraform code and identify potential issues and best practices violations. These tools can be integrated into the CI/CD pipeline to automatically detect and report issues before deploying infrastructure changes.

bashCopy code

```
# Running TFLint tflint --terraform-root=terraform-code
# Running Checkov checkov --directory=terraform-code
```

In these commands, TFLint and Checkov are executed against the Terraform codebase to identify linting issues and security misconfigurations. By incorporating these tools into the CI/CD pipeline, organizations can enforce coding standards, improve code quality, and mitigate security risks early in the development process.

Overall, automated testing and validation in infrastructure pipelines are essential for ensuring the reliability, security, and compliance of infrastructure deployments. By integrating testing frameworks, compliance tools, and linting utilities into the CI/CD workflow, organizations can achieve faster feedback loops, reduce manual effort, and improve the overall quality of their infrastructure code.

Chapter 6: Automating Multi-Account AWS Environments

Managing multiple AWS accounts with Terraform is a common requirement for organizations with complex infrastructure environments or those looking to enforce separation of concerns, security, and compliance. Terraform provides several features and best practices for effectively managing multiple AWS accounts, allowing users to provision, manage, and orchestrate infrastructure resources across different accounts seamlessly.

One of the fundamental concepts in managing multiple AWS accounts with Terraform is the use of AWS provider configurations. The AWS provider allows Terraform to interact with AWS services and resources, and it supports configuring multiple provider blocks to manage resources across different AWS accounts. Each provider block can specify different AWS credentials, regions, and account IDs, enabling Terraform to deploy resources to specific accounts.

hclCopy code

```
provider "aws" { region = "us-west-2" } provider "aws" { alias = "account_b" region = "us-east-1" access_key = "ACCOUNT_B_ACCESS_KEY" secret_key = "ACCOUNT_B_SECRET_KEY" }
```

In this example, two AWS provider blocks are defined: one for the default region in us-west-2 and another with an alias for a different account in us-east-1. The second provider block specifies custom access and secret keys for authenticating with the AWS account. By using provider aliases, Terraform can differentiate between multiple provider configurations and manage resources accordingly.

Another essential aspect of managing multiple AWS accounts with Terraform is the use of AWS IAM roles and cross-account access. IAM roles allow users or services in one AWS account to access resources in another account securely. Terraform supports assuming IAM roles programmatically, enabling users to delegate permissions across AWS accounts and provision resources using cross-account access.

hclCopy code

```
provider "aws" { alias = "account_b" region = "us-east-1"    assume_role    {    role_arn    = "arn:aws:iam::ACCOUNT_B_ID:role/TerraformRole" }
}
```

In this example, the AWS provider block is configured with an assume_role block specifying the ARN of an IAM role in another AWS account (account_b). When Terraform executes, it assumes the specified role in the target account, allowing it to provision resources using the permissions associated with that role.

Additionally, Terraform workspaces provide a mechanism for managing different environments or deployment stages within a single Terraform configuration. Workspaces allow users to maintain separate state files for each environment, enabling them to deploy the same infrastructure configuration to multiple AWS accounts or regions without interference.

bashCopy code

Creating a new workspace terraform workspace new dev # Selecting a workspace terraform workspace select dev

In these commands, Terraform creates a new workspace named 'dev' and selects it for use. Each workspace maintains its own state file, enabling users to manage resources independently across different AWS accounts or environments.

Moreover, Terraform's module composition capabilities enable users to define reusable infrastructure components and configurations that can be deployed across multiple AWS accounts. By encapsulating infrastructure logic and configurations into modules, users can promote consistency, reduce duplication, and simplify management across accounts.

hclCopy code

module "vpc" { source = "terraform-aws-modules/vpc/aws" version = "3.0.0" ... }

In this example, the 'terraform-aws-modules/vpc/aws' module is used to provision a VPC in the AWS account. The module abstracts the complexity of creating a VPC and provides configurable options for customization. By leveraging modules, users can standardize infrastructure patterns and deploy them consistently across different accounts.

Furthermore, Terraform's state management features, such as remote backends and locking mechanisms, are essential for managing infrastructure changes across multiple accounts securely. Remote backends store Terraform state files in a central location, facilitating collaboration and ensuring consistency across teams and environments.

hclCopy code

```
terraform { backend "s3" { bucket = "terraform-state"
key = "dev/terraform.tfstate" region = "us-west-2"
dynamodb_table = "terraform-lock" } }
```

In this configuration, Terraform is configured to store its state files in an S3 bucket with versioning enabled and uses a DynamoDB table for state locking. By centralizing state management, teams can avoid conflicts, track changes, and enforce access controls across multiple AWS accounts.

In summary, managing multiple AWS accounts with Terraform involves leveraging provider configurations, IAM roles, workspaces, modules, and state management features to orchestrate infrastructure resources effectively. By following best practices and utilizing Terraform's capabilities, organizations can achieve consistency, scalability, and security in managing infrastructure across complex AWS environments.

Implementing cross-account resource sharing with Terraform is a crucial aspect of managing complex cloud infrastructures where organizations need to securely share resources across different AWS accounts. This approach allows teams to maintain separation of concerns, enforce security boundaries, and facilitate collaboration while efficiently managing resources using Infrastructure as Code (IaC) principles. One common scenario for implementing cross-account resource sharing is sharing AWS IAM roles across accounts to grant access to specific resources or services. Terraform enables users to define IAM roles and policies within their infrastructure code and then assume these roles programmatically to access resources in different AWS accounts.

To implement cross-account IAM role sharing, users first define the IAM role and associated policies using Terraform's AWS provider:

```
hclCopy code
resource "aws_iam_role" "cross_account_role" {
name = "cross-account-role" assume_role_policy =
jsonencode({ Version = "2012-10-17" Statement = [ {
Effect = "Allow" Principal = { AWS =
"arn:aws:iam::ACCOUNT_ID:root" } Action =
"sts:AssumeRole" } ] }) } resource
"aws_iam_policy_attachment"
```

```
"cross_account_policy_attachment" { policy_arn =
aws_iam_policy.cross_account_policy.arn   roles   =
[aws_iam_role.cross_account_role.name] } resource
"aws_iam_policy" "cross_account_policy" { name =
"cross-account-policy" policy = jsonencode({ Version
= "2012-10-17" Statement = [ { Effect = "Allow"
Action    =    "s3:GetObject"    Resource    =
"arn:aws:s3:::example-bucket/*" } ] }) }
```

In this example, Terraform defines an IAM role
(**cross_account_role**) and an associated IAM policy
(**cross_account_policy**) that grants permissions to
access an S3 bucket (**example-bucket**). The
assume_role_policy specifies that the role can be
assumed by the root user of another AWS account
(**ACCOUNT_ID**).   Finally,   the
aws_iam_policy_attachment resource attaches the
policy to the IAM role.

Once the IAM role and policy are defined, users can
use Terraform's **assume_role** configuration block
within the AWS provider to assume the role
programmatically and access resources in the target
AWS account:

hclCopy code

```
provider "aws" { region = "us-west-2" assume_role {
role_arn                                              =
"arn:aws:iam::TARGET_ACCOUNT_ID:role/cross-
account-role" } }
```

In this configuration, Terraform's AWS provider is configured with an **assume_role** block specifying the ARN of the IAM role (**cross-account-role**) in the target AWS account (**TARGET_ACCOUNT_ID**). When Terraform executes, it automatically assumes the specified role, enabling it to provision resources in the target account.

Another common use case for cross-account resource sharing is sharing AWS VPC resources, such as subnets, route tables, and security groups, across multiple AWS accounts. Terraform allows users to define VPC resources in one account and then reference these resources in other accounts using data sources.

hclCopy code

```
data "aws_vpc" "shared_vpc" { id = "vpc-
1234567890" } resource "aws_subnet"
"subnet_in_other_account" { vpc_id =
data.aws_vpc.shared_vpc.id cidr_block =
"10.0.1.0/24" availability_zone = "us-west-2a" }
```

In this example, Terraform retrieves information about the shared VPC (**vpc-1234567890**) from the data source **aws_vpc**. Then, it creates a subnet (**subnet_in_other_account**) in a different AWS account using the VPC ID obtained from the data source.

Moreover, Terraform's support for remote backends and state sharing enables teams to collaborate on infrastructure configurations across multiple AWS accounts securely. By storing Terraform state files in a centralized location (e.g., Amazon S3 or Terraform Cloud), teams can share infrastructure code and coordinate changes seamlessly while maintaining isolation between accounts.

hclCopy code

terraform { backend "s3" { bucket = "terraform-state" key = "cross-account/terraform.tfstate" region = "us-west-2" dynamodb_table = "terraform-lock" } }

In this configuration, Terraform is configured to use an S3 backend to store its state files under the **cross-account** directory within the **terraform-state** bucket. Additionally, a DynamoDB table (**terraform-lock**) is used for state locking to prevent concurrent modifications.

Chapter 7: Advanced Load Balancing and Autoscaling Patterns

Advanced Load Balancer Configuration with Terraform offers powerful capabilities to optimize the performance, resilience, and security of applications deployed in cloud environments. By leveraging Terraform's declarative syntax and its integration with cloud provider APIs, developers and DevOps engineers can orchestrate sophisticated load balancer configurations with ease.

One of the key aspects of advanced load balancer configuration is the ability to fine-tune routing and traffic distribution strategies based on various criteria such as request attributes, geographic locations, or custom conditions. Terraform provides robust support for configuring these advanced routing rules using its AWS provider.

For example, to configure an Application Load Balancer (ALB) listener with path-based routing, one can use Terraform to define multiple listener rules that forward requests to different target groups based on the request path:

hclCopy code

```
resource "aws_lb_listener" "example" {
load_balancer_arn = aws_lb.example.arn port = 80
protocol = "HTTP" default_action { type = "forward"
target_group_arn = aws_lb_target_group.default.arn
```

```
} dynamic "default_action" { for_each =
var.listener_rules content { type = "forward"
target_group_arn =
default_action.value.target_group_arn order =
default_action.key + 1 dynamic "redirect" { for_each
= default_action.value.redirect != null ?
[default_action.value.redirect] : [] content {
status_code = redirect.value.status_code host =
redirect.value.host path = redirect.value.path query =
redirect.value.query } } } } }
```

In this example, Terraform dynamically generates ALB listener rules based on the input variable **listener_rules**, allowing users to define custom routing rules easily.

Additionally, Terraform enables users to configure advanced features of load balancers such as health checks, SSL termination, connection draining, and access logs. These features are essential for ensuring the reliability and security of applications running behind the load balancer.

hclCopy code

```
resource "aws_lb_target_group" "example" { name =
"example" port = 80 protocol = "HTTP" vpc_id =
aws_vpc.default.id health_check { path = "/" protocol
= "HTTP" timeout = 5 interval = 30 healthy_threshold
= 2 unhealthy_threshold = 2 matcher = "200" }
```

stickiness { type = "lb_cookie" cookie_duration = 3600 } tags = { Environment = "production" } }

In this snippet, Terraform defines an ALB target group with a custom health check configuration and sticky session settings, ensuring that traffic is routed only to healthy instances and that session affinity is maintained for a specified duration.

Furthermore, Terraform's support for infrastructure as code enables users to manage load balancer configurations alongside other infrastructure components, ensuring consistency and repeatability across environments.

hclCopy code

```
module "web_servers" { source = "terraform-aws-modules/ec2-instance/aws" count = var.instance_count name = "web-server-${count.index}" instance_type = var.instance_type ami = var.ami subnet_id = var.subnet_id key_name = var.key_name security_groups = [ aws_security_group.default.id, ] } resource "aws_lb_target_group_attachment" "example" { target_group_arn = aws_lb_target_group.example.arn target_id = module.web_servers.this_instance_id port = 80 }
```

In this example, Terraform provisions EC2 instances using a reusable module and attaches them to the previously defined target group, ensuring that traffic is properly routed to the instances by the load balancer.

Moreover, Terraform's support for versioning and state management enables teams to track changes to load balancer configurations over time and collaborate effectively on infrastructure changes.

In summary, advanced load balancer configuration with Terraform empowers users to implement sophisticated traffic routing, health monitoring, and security features for their applications with ease and consistency. By leveraging Terraform's declarative syntax and integration with cloud provider APIs, teams can efficiently manage complex load balancer configurations as part of their infrastructure as code workflows.

Dynamic autoscaling strategies with Terraform and AWS Auto Scaling provide organizations with the ability to automatically adjust the capacity of their infrastructure based on real-time demand, ensuring optimal performance, cost-efficiency, and resource utilization. Leveraging Terraform's infrastructure as code capabilities and AWS Auto Scaling's intelligent scaling policies, teams can implement sophisticated autoscaling strategies that respond dynamically to fluctuations in workload patterns.

To begin implementing dynamic autoscaling with Terraform and AWS Auto Scaling, the first step is to define the resources that will be autoscaled, such as EC2 instances or ECS tasks, using Terraform configuration files. For example, to create an Auto Scaling group for EC2 instances, one can use the following Terraform code:

hclCopy code

```
resource "aws_autoscaling_group" "example" { name = "example-asg" min_size = 1 max_size = 10 desired_capacity = 2 vpc_zone_identifier = [aws_subnet.example.id] launch_configuration = aws_launch_configuration.example.id tag { key = "Name" value = "example-instance" propagate_at_launch = true } }
```

In this configuration, Terraform defines an Auto Scaling group named "example-asg" with minimum and maximum sizes of 1 and 10 instances, respectively. The desired capacity is set to 2 instances, and the group is associated with a launch configuration named "example-instance."

Next, teams can define scaling policies to dynamically adjust the number of instances in response to changes in workload demand. AWS Auto Scaling supports various scaling policies, including target tracking, step scaling, and simple scaling. For example, to implement target tracking scaling based on CPU utilization, one can use Terraform to define a scaling policy as follows:

hclCopy code

```
resource "aws_autoscaling_policy" "cpu_scaling" { name = "cpu-scaling-policy" scaling_adjustment = 1
```

```
adjustment_type = "ChangeInCapacity" cooldown =
300              autoscaling_group_name        =
aws_autoscaling_group.example.name
target_tracking_configuration                    {
predefined_metric_specification                  {
predefined_metric_type                           =
"ASGAverageCPUUtilization" target_value = 50.0 } } }
```

In this example, Terraform creates a target tracking scaling policy named "cpu-scaling-policy" that adjusts the capacity of the Auto Scaling group based on the average CPU utilization of the instances, targeting a value of 50%. When the CPU utilization exceeds or falls below the target value, AWS Auto Scaling automatically adjusts the number of instances to maintain the desired level of utilization.

Additionally, teams can implement more advanced autoscaling strategies using custom CloudWatch metrics, such as application latency or queue length, to trigger scaling actions. By instrumenting applications to emit custom metrics and configuring CloudWatch alarms to monitor these metrics, teams can create highly responsive autoscaling policies that are tailored to the specific requirements of their applications.

Furthermore, Terraform's support for managing IAM roles and permissions enables teams to define granular access controls for autoscaling resources, ensuring that only authorized users can modify scaling policies and configurations.

hclCopy code

```
resource "aws_iam_role_policy_attachment"
"example" { role = aws_iam_role.example.name
policy_arn = aws_iam_policy.example.arn }
```

In this snippet, Terraform attaches an IAM policy to an IAM role, granting permissions to modify Auto Scaling resources.

Moreover, Terraform's plan and apply workflow provides visibility into the changes that will be made to the infrastructure before applying them, allowing teams to review and validate autoscaling configurations before deployment.

bashCopy code

```
terraform plan terraform apply
```

By following these best practices and leveraging Terraform's capabilities, organizations can implement dynamic autoscaling strategies with AWS Auto Scaling that effectively manage infrastructure capacity in response to changing workload demands, ensuring optimal performance, cost-efficiency, and reliability of their applications.

Chapter 8: Terraform Enterprise for Large-Scale Deployments

Deploying Terraform at enterprise scale involves implementing best practices, automation, and governance mechanisms to manage infrastructure efficiently and securely across large, complex organizations. At the core of this process is the use of Terraform Enterprise, which provides centralized control, collaboration features, and scalability necessary for managing infrastructure across multiple teams and environments.

To begin, organizations typically start by setting up Terraform Enterprise on-premises or using Terraform Cloud, which is a managed service offered by HashiCorp. This can be done by following the installation instructions provided by HashiCorp, which may involve setting up the necessary infrastructure, configuring authentication and access controls, and integrating with version control systems such as Git.

Once Terraform Enterprise is set up, organizations can establish a workflow for managing infrastructure as code (IaC) using Terraform configurations. This involves defining infrastructure resources, such as virtual machines, networks, and databases, using Terraform configuration files, which are then stored in version control repositories.

Teams can collaborate on infrastructure changes by using Terraform Enterprise's workspace feature, which allows multiple users to work on different branches of the same infrastructure configuration simultaneously. Workspaces provide isolation and versioning capabilities, enabling teams to experiment with changes in a safe and controlled manner.

As infrastructure configurations evolve, organizations can use Terraform Enterprise's policy as code feature to enforce compliance and security standards. Policies can be defined using Sentinel, a policy as code framework developed by HashiCorp, which allows organizations to define custom policies that are enforced during Terraform plan and apply operations.

In addition to policy enforcement, Terraform Enterprise provides visibility into infrastructure changes through its audit logging and monitoring features. Organizations can track who made changes to infrastructure configurations, when those changes were made, and the outcome of those changes, providing accountability and transparency.

To manage infrastructure at scale, organizations can leverage Terraform Enterprise's workspace management capabilities, which allow them to organize infrastructure configurations into logical groups based on projects, teams, or environments. Workspaces can be used to manage infrastructure across multiple clouds, regions, or accounts, enabling organizations to deploy and manage infrastructure consistently across their entire estate.

Furthermore, Terraform Enterprise provides integrations with popular CI/CD tools such as Jenkins, CircleCI, and GitLab CI/CD, enabling organizations to automate the deployment and testing of infrastructure changes as part of their continuous integration and delivery pipelines.

By deploying Terraform at enterprise scale, organizations can realize several benefits, including improved collaboration, increased agility, and reduced risk. However, deploying Terraform at scale also presents challenges, such as managing infrastructure drift, ensuring compliance, and scaling infrastructure management processes to accommodate growth.

To address these challenges, organizations should invest in training and education for their teams, establish clear governance processes, and leverage automation wherever possible. Additionally, organizations should stay abreast of new features and best practices in the Terraform ecosystem to ensure that they are maximizing the value of their investment in infrastructure as code.

In summary, deploying Terraform at enterprise scale requires careful planning, collaboration, and investment in tooling and processes. By following best practices and leveraging the capabilities of Terraform Enterprise, organizations can effectively manage infrastructure at scale while minimizing risk and maximizing efficiency.

Implementing collaborative workflows with Terraform Enterprise involves leveraging its features to enable multiple users to work together efficiently on infrastructure as code (IaC) projects. Terraform Enterprise provides a centralized platform for managing infrastructure configurations, version control integration, access controls, and collaboration tools, facilitating seamless collaboration among teams.

To begin, organizations typically set up Terraform Enterprise by deploying it on-premises or using Terraform Cloud, the managed service offered by HashiCorp. Installation and configuration involve setting up the necessary infrastructure, configuring authentication, integrating with version control systems like Git, and defining access controls to ensure that only authorized users can access and modify infrastructure configurations.

Once Terraform Enterprise is set up, teams can create workspaces for their projects. Workspaces serve as containers for organizing and managing infrastructure configurations, providing isolation, versioning, and collaboration capabilities. Creating a workspace can be done through the Terraform Enterprise UI or via Terraform CLI using the **terraform workspace** command.

bashCopy code

terraform workspace new <workspace_name>

Teams can then define their infrastructure using Terraform configuration files, which describe the desired state of the infrastructure resources. These configuration files are stored in version control repositories connected to Terraform Enterprise, allowing teams to track changes, review code, and collaborate on infrastructure changes using familiar version control workflows.

As multiple users collaborate on infrastructure changes, Terraform Enterprise's locking mechanism ensures that only one user can apply changes to a workspace at a time, preventing conflicts and ensuring consistency. Users can acquire a lock using the Terraform CLI **terraform apply** command.

bashCopy code

terraform apply -lock=true

Terraform Enterprise also provides features for code review and collaboration, allowing teams to review infrastructure changes before they are applied. Users can submit changes for review using pull/merge requests in the version control system, and reviewers can provide feedback and approve changes directly within the Terraform Enterprise UI.

Once changes are approved, teams can apply them to the infrastructure using Terraform Enterprise's run environment. Terraform runs can be triggered manually through the UI or automatically triggered by events such as code commits or pull request merges. The **terraform apply** command is used to apply changes to the infrastructure.

bashCopy code

terraform apply

During the apply process, Terraform Enterprise provides visibility into the execution plan, showing the changes that will be made to the infrastructure resources. This allows users to review and approve the changes before they are applied, ensuring that only authorized changes are made to the infrastructure.

In addition to manual runs, Terraform Enterprise supports automated runs through its API, enabling integration with CI/CD pipelines. Teams can trigger

Terraform runs as part of their continuous integration and delivery workflows, ensuring that infrastructure changes are tested and deployed automatically.

Terraform Enterprise also provides audit logging and monitoring capabilities, allowing organizations to track who made changes to the infrastructure, when those changes were made, and the outcome of those changes. This provides visibility and accountability, helping organizations meet compliance and security requirements.

Overall, implementing collaborative workflows with Terraform Enterprise enables teams to work together efficiently on infrastructure projects, ensuring consistency, reliability, and security. By leveraging Terraform Enterprise's features for version control integration, access controls, code review, and automation, organizations can streamline their infrastructure deployment processes and accelerate their time to market.

Chapter 9: Managing Secrets and Sensitive Data in Terraform

Implementing secure secret management with Terraform involves leveraging its capabilities to store and manage sensitive information such as API keys, passwords, and other credentials securely. Securing secrets is crucial for protecting sensitive data and preventing unauthorized access to critical resources. Terraform provides several techniques and best practices for managing secrets securely, including the use of state encryption, environment variables, external vault integration, and provider-specific secret management solutions.

One approach to secure secret management in Terraform is to encrypt the state file, which contains sensitive information about the infrastructure configuration. Terraform Enterprise and Terraform Cloud offer built-in state encryption features that encrypt the state file at rest, ensuring that sensitive data is protected from unauthorized access. To enable state encryption in Terraform Cloud, users can enable the "State Versioning" and "State Encryption" settings in the workspace configuration.

bashCopy code

```
terraform workspace select <workspace_name>
terraform state push
```

Another technique for managing secrets securely is to use environment variables to pass sensitive information to Terraform during runtime. Environment variables can be set locally or in CI/CD pipelines to provide credentials and other sensitive data to Terraform without exposing them in plain text. Terraform automatically reads environment variables prefixed with **TF_VAR_** and uses them as input variables in the configuration.

bashCopy code

```
export TF_VAR_secret_key="my_secret_key"
```

Additionally, organizations can integrate Terraform with external secrets management systems such as HashiCorp Vault, AWS Secrets Manager, or Azure Key Vault to centralize and secure secret storage. These solutions provide robust encryption, access control, and auditing capabilities for managing secrets at scale. Terraform supports integration with these external vaults through provider plugins, allowing users to retrieve and use secrets dynamically in their configurations.

bashCopy code

```
terraform { required_providers { vault = { source = "hashicorp/vault" version = ">= 2.0" } } } provider "vault" { address = "https://vault.example.com" token = "vault_token" } data "vault_generic_secret" "example" { path = "secret/data/example" }
```

Furthermore, cloud providers offer native secret management solutions that integrate seamlessly with Terraform. For example, AWS Secrets Manager and Azure Key Vault allow users to store and manage secrets securely within their respective cloud platforms. Terraform provides provider-specific resources for interacting with these services, enabling users to create, retrieve, and manage secrets directly from their Terraform configurations.

hclCopy code

```
resource "aws_secretsmanager_secret" "example" {
name = "example-secret" } resource
"aws_secretsmanager_secret_version" "example" {
secret_id = aws_secretsmanager_secret.example.id
secret_string = "super_secret_value" }
```

Additionally, Terraform modules can be used to encapsulate secret management logic and enforce best practices across different projects and teams. By creating reusable modules for secret management, organizations can standardize their approach to security and ensure consistency in how secrets are managed and accessed across their infrastructure.

In summary, implementing secure secret management with Terraform is essential for protecting sensitive data and ensuring the security of infrastructure deployments. By leveraging features such as state encryption, environment variables, external vault integration, and provider-specific secret management solutions, organizations can effectively manage secrets and mitigate the risk of unauthorized access to critical resources.

Integrating with AWS Key Management Service (KMS) for encryption is a crucial aspect of securing data in AWS environments. AWS KMS is a fully managed encryption service that allows users to create and control cryptographic keys used to encrypt data. By leveraging AWS KMS, organizations can encrypt sensitive data stored in various AWS services such as Amazon S3, Amazon EBS, and Amazon RDS, ensuring that data remains confidential and protected from unauthorized access.

To integrate with AWS KMS for encryption, users first need to create a Customer Master Key (CMK) in the AWS

Management Console or via the AWS CLI. The CMK serves as the root key for encrypting and decrypting data and can be either customer-managed or AWS-managed. Creating a CMK can be done using the AWS CLI as follows:
bashCopy code

```
aws kms create-key --description "My CMK"
```

Once the CMK is created, users can specify it when encrypting data using AWS services or client-side encryption libraries. For example, when uploading an object to Amazon S3, users can specify the CMK to use for encryption:
bashCopy code

```
aws s3 cp my_file.txt s3://my_bucket/ --sse aws:kms --sse-kms-key-id <KMS_key_id>
```

Alternatively, users can use the AWS SDKs or APIs to encrypt data programmatically, specifying the CMK to use for encryption. For example, in the AWS SDK for Python (Boto3), users can encrypt data using a specific CMK:
pythonCopy code

```
import boto3 kms = boto3.client('kms') response = kms.encrypt( KeyId='<KMS_key_id>', Plaintext=b'Hello, World!' ) print(response['CiphertextBlob'])
```

AWS KMS also supports envelope encryption, where data is encrypted using a data encryption key (DEK), and the DEK is then encrypted using the CMK. This allows users to encrypt large volumes of data efficiently while still benefiting from the security provided by AWS KMS. Envelope encryption can be used with various AWS services, including Amazon S3, Amazon EBS, and Amazon RDS.

In addition to encrypting data at rest, AWS KMS can also be used to encrypt data in transit. For example, users can use AWS KMS to encrypt messages sent between services using Amazon Simple Queue Service (SQS) or Amazon Simple Notification Service (SNS), ensuring that data remains encrypted while in transit.

AWS KMS provides granular access control through key policies and IAM policies, allowing users to define who can use and manage encryption keys. Key policies specify permissions for the CMK, such as who can encrypt or decrypt data using the key, while IAM policies control access to AWS KMS APIs and resources.

Furthermore, AWS KMS integrates seamlessly with other AWS services, such as AWS CloudTrail and AWS CloudWatch, providing logging and monitoring capabilities for key usage and security auditing. Users can monitor key usage and access patterns to detect any unauthorized or suspicious activities.

Overall, integrating with AWS KMS for encryption is a critical aspect of securing data in AWS environments. By leveraging AWS KMS, organizations can encrypt sensitive data at rest and in transit, control access to encryption keys, and monitor key usage for compliance and security purposes.

Chapter 10: Real-Time Monitoring and Observability with Terraform

Setting up real-time monitoring solutions with Terraform is crucial for organizations to proactively monitor their infrastructure and applications, identify issues promptly, and ensure optimal performance and reliability. Real-time monitoring allows for the continuous collection and analysis of metrics, logs, and events from various sources, providing insights into the health and performance of the entire system.

One essential component of setting up real-time monitoring solutions is the deployment of monitoring agents or collectors on the target infrastructure. These agents are responsible for gathering metrics and logs from the system and forwarding them to the monitoring platform. Terraform can automate the deployment of monitoring agents using infrastructure as code (IaC) principles, ensuring consistency and repeatability across environments.

For example, to deploy the Datadog agent on an EC2 instance using Terraform, users can define an EC2 instance resource along with user data that installs and configures the Datadog agent during instance bootstrapping:

hclCopy code

resource "aws_instance" "example" { ami = "ami-12345678" instance_type = "t2.micro" user_data = <<-EOF #!/bin/bash sudo apt-get update sudo apt-get install -y datadog-agent sudo systemctl enable datadog-agent sudo systemctl start datadog-agent EOF }

This Terraform configuration launches an EC2 instance and automatically installs and starts the Datadog agent upon instance creation. Similar configurations can be created for other monitoring solutions such as Prometheus, Grafana, or New Relic.

Once monitoring agents are deployed, the next step is to configure them to collect relevant metrics, logs, and events. This involves defining monitoring checks, alerting rules, and dashboards tailored to the specific needs of the application or infrastructure being monitored. Terraform can manage these configurations using provider-specific resources or by executing external scripts or configuration management tools.

For example, to configure monitoring checks and alerting rules in Datadog using Terraform, users can define resources that represent monitors and alerting policies:

hclCopy code

resource "datadog_monitor" "high_cpu_usage" { name = "High CPU Usage" type = "metric alert" query = "avg(last_5m):avg:system.cpu.user{host:host0} > 90" message = "High CPU usage detected on host0" escalation_message = "High CPU usage persists on host0" notify_no_data = false notify_audit = false timeout_h = 1 thresholds { critical = 90 } } resource "datadog_monitor" "disk_space" { name = "Low Disk Space" type = "service check" query =

"service:disk_space.dev./.percentage_used > 80" message = "Low disk space detected on /" escalation_message = "Low disk space persists on /" notify_no_data = false notify_audit = false timeout_h = 1 }

In this example, two Datadog monitors are defined— one for high CPU usage and another for low disk space. These monitors trigger alerts when the specified conditions are met, notifying the appropriate stakeholders via email, Slack, or other channels.

Additionally, Terraform can be used to deploy and configure real-time logging solutions, such as the Elastic Stack (Elasticsearch, Logstash, and Kibana), Fluentd, or AWS CloudWatch Logs. These solutions aggregate and analyze logs from various sources, allowing users to gain insights into application and system behavior, troubleshoot issues, and meet compliance requirements.

For example, to deploy an Elasticsearch cluster with Terraform, users can define resources that provision the necessary AWS infrastructure (e.g., EC2 instances, security groups, and IAM roles) and install Elasticsearch using user data or custom AMIs.

hclCopy code

```
resource "aws_instance" "elasticsearch" { ami = "ami-12345678" instance_type = "t2.medium" subnet_id = aws_subnet.example.id security_group_ids = [aws_security_group.example.id] iam_instance_profile = aws_iam_instance_profile.example.name user_data =
```

```
<<-EOF #!/bin/bash sudo apt-get update sudo apt-get
install -y openjdk-11-jre wget -qO -
https://artifacts.elastic.co/GPG-KEY-elasticsearch |
sudo apt-key add - sudo apt-get install -y apt-transport-
https echo "deb
https://artifacts.elastic.co/packages/7.x/apt stable
main" | sudo tee -a /etc/apt/sources.list.d/elastic-
7.x.list sudo apt-get update sudo apt-get install -y
elasticsearch sudo systemctl enable elasticsearch sudo
systemctl start elasticsearch EOF }
```

This Terraform configuration deploys an EC2 instance with Elasticsearch installed, allowing users to store and analyze logs in real-time. Similar configurations can be created for Logstash and Kibana to complete the Elastic Stack deployment. Implementing observability practices for infrastructure as code (IaC) is crucial for ensuring the reliability, performance, and security of modern cloud-based environments. Observability refers to the ability to understand the internal state of a system based on its external outputs. In the context of IaC, observability encompasses monitoring, logging, tracing, and alerting mechanisms that provide insights into the health and behavior of infrastructure resources and applications deployed using code.

One fundamental aspect of implementing observability practices for IaC is setting up comprehensive monitoring solutions. Monitoring involves collecting and analyzing metrics and performance data from various components of the infrastructure to assess its health and performance continuously. Terraform, a popular IaC tool, can be used to deploy and configure monitoring infrastructure and agents seamlessly.

To set up monitoring using Terraform, users can leverage provider-specific resources to provision monitoring services such as Amazon CloudWatch, Datadog, Prometheus, or Grafana. For instance, to deploy CloudWatch alarms for monitoring EC2 instance CPU utilization, users can define CloudWatch metric alarm resources in their Terraform configuration:

hclCopy code

```
resource "aws_cloudwatch_metric_alarm" "cpu_utilization_alarm" { alarm_name = "HighCPUUtilization" comparison_operator = "GreaterThanThreshold" evaluation_periods = "2" metric_name = "CPUUtilization" namespace = "AWS/EC2" period = "120" statistic = "Average" threshold = "70" alarm_description = "Alarm when CPU utilization exceeds 70%" dimensions = { InstanceId = aws_instance.example.id } alarm_actions = [aws_sns_topic.example.arn] }
```

In this example, Terraform is used to create a CloudWatch alarm that monitors the CPU utilization of an EC2 instance and triggers an SNS notification when it exceeds the specified threshold.

Another essential aspect of observability is logging, which involves capturing, storing, and analyzing log data generated by infrastructure resources and applications. Terraform can automate the deployment and configuration of logging solutions such as the Elastic

Stack (Elasticsearch, Logstash, and Kibana), Fluentd, or AWS CloudWatch Logs.

For instance, to deploy an Elasticsearch cluster for centralized logging with Terraform, users can define resources that provision the necessary infrastructure components and install Elasticsearch using user data or custom AMIs, similar to the example provided earlier for monitoring.

Additionally, Terraform can be used to configure log forwarding agents or integrations on target resources to stream logs to the central logging solution. For example, to configure Fluentd on an EC2 instance to forward logs to an Elasticsearch cluster, users can define a user data script that installs and configures Fluentd accordingly.

hclCopy code

```
resource "aws_instance" "example" { ami = "ami-12345678" instance_type = "t2.micro" user_data = <<-EOF #!/bin/bash # Install Fluentd curl -L https://toolbelt.treasuredata.com/sh/install-ubuntu-xenial-td-agent3.sh | sh # Configure Fluentd to forward logs to Elasticsearch echo ' <match **> @type elasticsearch host <elasticsearch-host> port 9200 logstash_format true </match>' | sudo tee -a /etc/td-agent/td-agent.conf # Restart Fluentd sudo systemctl restart td-agent EOF }
```

This Terraform configuration deploys an EC2 instance with Fluentd installed and configured to forward logs to an Elasticsearch cluster.

In addition to monitoring and logging, implementing observability practices for IaC also involves setting up distributed tracing solutions to track and analyze requests as they flow through complex distributed systems. Solutions such as AWS X-Ray, Jaeger, or Zipkin can be integrated into IaC workflows using Terraform to enable distributed tracing capabilities.

To integrate AWS X-Ray with Terraform, users can define resources that enable X-Ray tracing for AWS Lambda functions, API Gateway endpoints, or EC2 instances, depending on the application architecture.

hclCopy code

```
resource "aws_xray_sampling_rule" "example" {
service_name = "my-service" service_type =
"AWS::Lambda::Function" http_method = "*" url_path
= "*" fixed_rate = 0.1 }
```

This Terraform configuration creates an AWS X-Ray sampling rule that applies to AWS Lambda functions with a fixed sampling rate of 10%.

In summary, implementing observability practices for infrastructure as code involves setting up comprehensive monitoring, logging, and tracing solutions using tools like Terraform to automate the deployment and configuration of monitoring infrastructure, agents, and integrations. By leveraging observability practices, organizations can gain valuable insights into their cloud-based environments, detect and diagnose issues promptly, and ensure the reliability and performance of their infrastructure and applications.

Conclusion

In summary, the book bundle "Harnessing Terraform for AWS Infrastructure as Code" offers a comprehensive journey through the world of cloud automation and infrastructure management using Terraform. Across the four books included in this bundle, readers are equipped with a holistic understanding of Terraform's capabilities, from essential concepts for beginners to advanced techniques and strategies for optimizing and managing complex AWS deployments.

In "AWS Cloud Automation: Terraform Essentials for Beginners," readers are introduced to the fundamental concepts of Terraform and its role in automating infrastructure provisioning on AWS. They learn how to define infrastructure as code, manage resources using Terraform configurations, and deploy basic AWS environments efficiently.

"Mastering Terraform: Advanced Techniques for AWS Cloud Automation" delves into more advanced topics, providing readers with in-depth insights into Terraform's advanced features and best practices. From managing state and dependencies to implementing modularization and reusable modules, readers gain the expertise needed to tackle complex infrastructure automation projects with confidence.

"Optimizing AWS Infrastructure: Advanced Terraform Strategies" takes readers on a journey toward maximizing the efficiency and performance of AWS infrastructure deployments using Terraform. Through optimization techniques, readers learn how to minimize costs, enhance scalability, and improve resource utilization, ensuring their infrastructure meets evolving business requirements.

Finally, "Expert AWS Cloud Automation: Scaling and Managing Complex Deployments with Terraform" offers an advanced exploration of Terraform's capabilities for scaling and managing complex AWS deployments. Readers learn how to orchestrate multi-region architectures, implement advanced networking configurations, and handle sophisticated deployment workflows with ease.

Collectively, these four books provide a comprehensive resource for individuals and teams looking to harness the power of Terraform for AWS infrastructure as code. Whether you're a beginner looking to get started or an experienced practitioner seeking to refine your skills, this bundle equips you with the knowledge and techniques needed to succeed in the world of cloud automation and infrastructure management.

www.ingramcontent.com/pod-product-compliance
Lightning Source LLC
Chambersburg PA
CBHW070935050326
40689CB00014B/3208